Android® SmartPhones For Seniors

by Marsha Collier

for dummies®
A Wiley Brand

Android® Smartphones For Seniors For Dummies®

Published by: **John Wiley & Sons, Inc.**, 111 River Street, Hoboken, NJ 07030-5774, www.wiley.com

Copyright © 2022 by John Wiley & Sons, Inc., Hoboken, New Jersey

Published simultaneously in Canada

For general information on our other products and services, please contact our Customer Care Department within the U.S. at 877-762-2974, outside the U.S. at 317-572-3993, or fax 317-572-4002. For technical support, please visit https://hub.wiley.com/community/support/dummies.

Wiley publishes in a variety of print and electronic formats and by print-on-demand. Some material included with standard print versions of this book may not be included in e-books or in print-on-demand. If this book refers to media such as a CD or DVD that is not included in the version you purchased, you may download this material at http://booksupport.wiley.com. For more information about Wiley products, visit www.wiley.com.

Library of Congress Control Number: 2021947994

ISBN 978-1-119-82848-8 (pbk); ISBN 978-1-119-82849-5 (ePDF); ISBN 978-1-119-82850-1 (epub)

Contents at a Glance

Table of Contents

Introduction

This book is for you, even if you're not a senior. You no doubt live a busy life and may rarely have the desire (or time) to learn to operate every last feature of your Android smartphone. Android has *so* many options, and in this book, I share some of my favorites, some hidden nuggets, and some basic tricks for beginning (and not-so-beginning) users.

I've met too many people who refuse to learn anything regarding technology, and it upsets me. As you grow older (don't we all?), improving cognitive abilities is a top way to fend off mental decline. What better way to boost brainpower than to master the tiny computer in your pocket? Challenge yourself — think of spending time on your smartphone as a form of self-care (and having fun at the same time).

Hopefully, this book can keep you from making embarrassing calls to your kids to "fix" your phone. And even better — this book might help you teach your kids a trick or two. My editors don't necessarily identify with the over 55 crowd, but even they found some good ideas in this book.

So, buy the book. I know we'll have fun learning together.

About This Book

The idea for this book was born during the coronavirus pandemic, on the day that vaccinations became available in Los Angeles. I scoured the Internet, found the active link from the misleading information, and wanted to be sure that my over-65 neighbors could sign up for the vaccine as well.

I went from house to house, helping others sign up for an appointment (much to their relief). The problem was not necessarily determining where to go for shots, but rather how to schedule the appointments on their phones.

The over 55 crowd had problems with both Apple and Android phones. Although I've had short stints with iPhones, I am a loyal Android user. I prefer the interface, I love the idea that I can personalize what information I share (and with who), and most of all, because "Android is for everyone."

From my research, I've found that not everyone rushes out to buy the latest, most expensive phones. I suspected as much because that's my philosophy as well. So, I used six different phone models and brands, and not necessarily the latest-and-greatest. The introduction's figure shows some of my "coauthors" for this book.

The Coauthors

The coauthor phones ran the latest version of Android, 11. They include a 2021 OnePlus 9 Pro 5G, a late 2019 Samsung Galaxy Note10+, 2019 Samsung Galaxy S10+, 2021 Samsung Galaxy A32 5G, 2018 Google Pixel 3XL, and a 2017 Pixel 2. I also checked the between Android versions on a 2019 Huawei P30 Pro running Android 10. Using all these phones was confusing at worst, but at best, I learned that the Android experience flows similarly through all the brands, with additional features, which I'll show you in the book.

All these phones are *impressive*, and the manufacturers did a great job with the design and functionality. The oldest, the 2018 Google Pixel 3XL, is quite a surprise; I enjoyed using that phone as well.

I wrote this book using the pictured phones, all but one of which was running the Android 11 operating system. Android 12 launches in late 2021, so I used a OnePlus 9 5G on the beta tests of Android 12. Chapter 17 has my impressions on the beta-testing experience.

Foolish Assumptions

I assume that you have used a smartphone — or at least used a computer. I also assume that you may be in the market for a phone or have just acquired one. I guide you through the intricacies of buying a phone that best meets your needs, the specific features to look for in a mobile service provider, the basic operations of navigating your Android smartphone, and the most important apps you need to master (even for photography and entertainment). I highlight ways to use your phone that we (my editors and I) expect will improve your Android experience. Most importantly, I present the content in nontechnical language!

Icons Used in This Book

I'm all about illustrating a book. Many figures illustrate the phone screens and tasks laid out in the chapters. You'll also see small icons that point to special advice from me — here they are:

TIP

Tips are short notes — from me to you — that make the process of using your smartphone easier.

TECHNICAL
STUFF

These notes are like tips, but a bit more technical sounding. I provide only the technical info that's both important and worthwhile to know.

REMEMBER

Don't forget these notes. They show up often and are good to stash in the back of your mind.

WARNING

Warnings contain information that you need to heed. Sometimes, performing certain actions can cause an expensive problem, but *don't worry:* You won't find many of these.

Beyond the Book

You can find a section on www.dummies.com with more Android tips and Android 12 features (after the Android 12 operating system fully rolls out). While visiting dummies.com, type **Android Smartphones For Seniors For Dummies** in the search box to find the book's Cheat Sheet. I also have a blog at https://mcollier.blogspot.com, where I post related articles.

In this book, I give you links and contact information for the tech support department for several major phone manufacturers, which should help with any immediate issue.

If you have specific questions, feel free to reach me by completing the contact form at my website (www.coolebaytools.com/contact) or on any of the social media platforms. I'd love to hear from you about the topics you want me to add to the next edition.

I also have a podcast, "Computer and Technology Radio," which I host with Marc Cohen. You can find it on your favorite podcast platform.

Where to Go from Here

Start reading this book anywhere you want. An incredible index can help you find almost any topic you need to learn about. Go to the table of contents and pick a chapter that interests you, or read the book from the beginning.

Interesting and helpful nuggets of information for you abound every-where in this book. I really hope you enjoy it.

1

Your Phone in the Android Universe

Chapter **1**

Why Android? What's the Deal?

The smartphone is undoubtedly the most common yet powerful personal technology in your life. In the United States market, you have basically this choice: an Apple iPhone (iOS) or an Android-based phone.

The smartphone platform you choose is a matter of preference. Some people use both Apple and Android products, but in the end usually tend to favor one platform over the other.

Because the iPhone lives in a *walled garden,* Apple makes the decisions and takes the profits on the phone, accessories, services, and apps. Apple products and software work in sync; you have few choices to make.

Conversely, no matter the brand, all Android phones have similar genetics and are the same at their core. But you find a variety of options for accessories, phone brands, services, apps, and (most of all) prices.

The competition for dollars in the Android marketplace begets innovation, and I believe that opting to spend *your* dollars in that marketplace is a good choice. In this chapter, I offer foundational information to help make the Android experience even better.

A Little Android History

Android is the operating system on the majority of the world's smartphones. It's an open source operating system led by Google (according to `https://source.android.com`):

> As an open source project, Android's goal is to avoid any central point of failure in which one industry player can restrict or control the innovations of any other player.

Translation: If you come up with a device, you too can use the Android operating system to power it for free. (I'm not that smart.)

I believe that much of the magic of Android lies with Google Mobile Services (GMS) — the collection of apps and functionalities that make the Android ecosystem a *useful* environment.

As an Android user, you may wonder why you feel more comfortable using the operating system. The answer is a surprise to many: the Android mobile operating system is based on Linux (another open source operating system) and many Microsoft patents. The influence of these two giants in the software world accounts for the familiarity, and thus the comfort, you may feel when using an Android device. The nearby sidebar "The scope of Microsoft involvement" gives a quick look at the Android–Microsoft connection.

TIP

Note that many Android phones come with Microsoft Office pre-installed. Also, did you know that you can send text messages on your phone from your Windows desktop PC? Android phone owners can just go to `messages.google.com/web` to connect the devices.

The Many Flavors (Versions) of Android

Throughout this book, I offer stories about the beginnings of, and evolution of, the Android platform. For now, you should know that the version of Android I'm using for this book is Android 11. Chapter 17 talks about Android 12, which is scheduled to release around the time this book publishes. I got hold of a beta (not-ready-for-prime-time) version so that I can see the similarities and differences in the new upgrade.

The Android mascot (Bugdroid) is a small, green robot, shown on the left in **Figure 1-1**. Bugdroid, who gets dressed up with each new version of Android, was designed in 2008 to be an internationally understood symbol — like airport signs — because Android was designed for everyone. Since the platform's inception, Bugdroid appears in advertising and has undergone minor changes over the years. Today, the mascot's green color is updated, and just the top of the head shows (on the right in Figure 1-1).

"The Android robot is reproduced from work created and shared by Google and used according to terms described in the Creative Commons 3.0 Attribution License." https://source.android.com/setup/start/brands#robot-android

FIGURE 1-1

The initial versions of Android, except for A and B, had nicknames based on sweet treats. The nicknames gave a friendly, homey feeling to each update and personalized it to its users. Android 1.5 (Android C) was Cupcake, which was apparently when the naming system began. From there on, the charming version nicknames followed the alphabet, and even though Google publicly discontinued the cute code nicknames, Android 11 (Android R) is Red Velvet Cake, and (rumor has it), Android 12 (Android S) is Snow Cone.

Reasons That People Choose Android

REMEMBER

Not all smartphones are iPhones.

The most excellent aspect of Android is that it's customizable. If you don't like one way of doing things, you can switch to another mode. Android is as simple or as complex as you want it to be.

Also, because Android is a free, open source platform, you can buy Android phones at many price points. You can even buy a brand-new phone inexpensively. Samsung has an A series (*A* for *affordable,* perhaps?) and other manufacturers make basic phones that won't put a dent in your budget. The more bells and whistles a manufacturer adds, the more it ratchets up the price point. A *flagship* (top of the line) Android phone can be as expensive as any other. (Find more advice about picking out your phone in Chapter 2.)

Here are several features I love about owning an Android smartphone:

» **Keyboards and default apps:** I love having the option to try out different keyboards and browsers. You can download many new apps for free from the Google Play Store and try them out. If you come to realize that an app isn't your cup of tea, just uninstall it.

» **Sharing:** Whenever you want to share a photo, a web page, an email — you name it — tapping the Share icon brings up a simple sheet with app icons. Just tap the one you want to share to, and you're on the way.

» **Navigating screens:** As many times as I've used iOS devices (iPhone, iPad), my productivity always freezes when it comes to going back a page or a screen. On Android, you can use hand gestures or the bottom-of-screen Back arrow to navigate simply.

» **Notifications:** They're easy to control. You can control how you see them organized on the Home screen, app by app.

These are just a few features, but the truth is, Android is about choice. You're not forced to use any specific brand, and your phone can link to many different devices in your home (such as the thermostat, electric outlets, or lights).

Why You Need a Google Account

If you've ever purchased an Apple product (iPhone, iPod, or iPad), you know that you had to sign up for an Apple ID. Apple asked me to input my credit card information, even though I had no desire to purchase anything in the App Store. I couldn't register the phone without it.

To use the Google apps, you also need an ID, which is officially your account. *But* you don't have to supply any credit card information until you reach the point where you actually want to purchase something.

REMEMBER

Technically speaking, you don't *need* a Google account, but I believe that it makes your Android experience better. A Google account is required in order to identify you over Android Mobile Services. You need the account to gain full value from the many native Android apps, such as Gmail, Calendar, Photos, Play Store, or Maps.

Accessing apps and settings

The native apps are free to use, but accessing everything from your account, all in one place, is *handy*. And all the apps are personalized just for you, based on how you use each app on the platform.

You can access your account on your devices (or laptop) by going to `myaccount.google.com`. From this screen (shown on the left in **Figure 1-2**), you can adjust the settings related to your interactions with Google. The ability to edit settings related to everything from your personal information to privacy, security, and more appears on the tabs on these screens.

TIP

If your photo doesn't appear in your Google Account, tap the letter that's in the photo circle to upload one from your device or PC. That way, whenever you are in a Google app on a laptop or on your phone, your picture appears in the upper right. Tap it, and it opens up Manage Your Google Account.

Bequeathing your account

From your Google account, you can determine what should be done with the account should you not log in for a while. Google can notify someone you name, give that person access to your data, or delete the account altogether (refer to the right side of Figure 1-2). I definitely want my daughter to have access to my photo archive after I'm gone. Anyway, you can set it all up there.

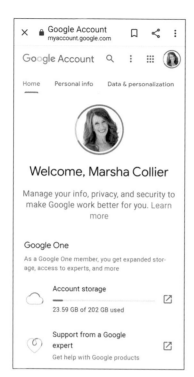
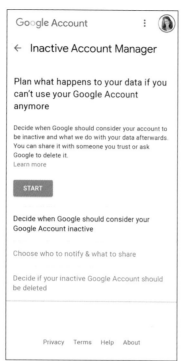

FIGURE 1-2

GOOGLE ONE

After you've used your Google account for a while — and backed up your data — you may fill up a lot of storage space in your complimentary 15 GB or, *gigabytes*. (One GB is approximately 1,000 MB, or *megabytes*).

Everything you save on Google counts toward this storage amount — Gmail messages, photos, documents in Drive. I've had my Gmail account for years and use it as a *de facto* filing cabinet. I've saved all my important emails there because they're easy to search; it's a perfect system. I uploaded my entire photo library to Google Photos, not only for archival purposes but also because Google Photos has fun and useful features. Going beyond a digital storehouse, Google Photos enables you to print photo books with selected pictures to give as special-occasion gifts.

After almost a decade, I have exceeded my 15 free gigabytes and now need to pay for more storage. The price was quite reasonable. I signed up for 200GB of

(continued)

(continued)

storage for $29.99 a year — a small price to pay for the security of having my data backed up.

With the new storage on hand, I now have a Google One account, which provides the aforementioned 200GB of storage, access to Google experts for premium support for any problems, 3 percent back in credit at the online Google store (https://store.google.com), and more.

To find out more about Google One services, download the Google One app from the Play Store or go to one.google.com in your browser on your phone or laptop.

So Many Choices!

Many companies make Android phones, and each one works hard to put its own spin on the device. Or not. A Google Pixel phone is pure Android. For those who remember the Blackberry, I've even heard rumors that a new Blackberry model based on Android is coming out!

The additions that manufacturers put on their devices become Android combined operating systems, or *skins*, as software overlays that deliver the interface design of the phone. A lot of extra software can be piled on a phone this way. I remember being given a popular brand's phone to review, and I couldn't get past the complexity of the skin. Screen shots in this book illustrate that similar screens can look different between phones because each manufacturer's device has its own skin design. You'll notice that even Android icons may appear in circles or squares and are modified ever so slightly.

Android skin customizations offer the user an enhanced experience. Its up to you to decide which one works best in your day-to-day phone use.

TECHNICAL STUFF

Just as Android names each version of the software, it's probably no surprise (because branding is important to manufacturers) that each OS skin has its own name: OncPlus = OxygenOS, Samsung = One UI, Xiaomi = MIUI, Oppo = ColorOS.

Note: You receive separate updates on your phone for security, the Android operating system, and the manufacturer's operating system.

Tech support options

In Chapter 2, you find out about buying a phone, but know that you don't have to buy from the cellular carrier. You can buy direct. Because both the manufacturer and carrier are intertwined with your phone, you *might* receive better tech support by contacting the manufacturer.

In **Table 1-1**, I give you tech support contact information for the major phone manufacturers. I have added a Twitter handle for brands that have a dedicated tech support Twitter account. Reaching brands via social media can be the fastest way to find service.

REMEMBER

Yes, efficient service *can* happen! On a personal note, I can attest to reaching Samsung support on Twitter (some years ago). Even though my device was out of warranty and the problem was caused by my own idiocy (I put the phone on a counter and spilled a cleaning solution which saturated my phone), I told them I'd pay for the repair. They asked that I mail in my phone, and I was able to track the repair progress on the online portal. The process was super efficient, and my phone was returned to me quickly.

TABLE 1-1 **Customer Service Contact Info for Android Phone Manufacturers**

Manufacturer	Twitter Handle	Link	Phone
Huawei		`https://consumer.huawei.com/us`	(888) 548-2934
Lenovo	@LenovoSupport	`https://support.lenovo.com/`	(866) 426-0911
LG	@LGUSSupport	`https://www.lg.com/us/mobile`	(800) 243-0000
Motorola	@Moto_Support	`https://motorola-global-portal.custhelp.com/app/mcp/contactus`	(800) 734-5870

(continued)

TABLE 1-1 *(continued)*

Manufacturer	Twitter Handle	Link	Phone
OnePlus	@OnePlus_Support	`https://www.oneplus.com/support`	(833) 777-3633
Oppo	@OPPOCareGlobal	`https://support.oppo.com/en`	(800) 103-2777
Pixel	@Google	`https://support.google.com/pixelphone`	
Samsung	@SamsungSupport	`https://www.samsung.com/us/support`	Text HELP to 62913
TCL	@TCL_USA	`https://support.tcl.com`	(855) 224-4228
Xiaomi		`https://www.mi.com/global`	(833)-942-6648
ZTE		`https://ztedevices.com/en-gl/support-center`	(877) 817-1759

5G? LTE? 4G? VoLTE? Whaaat?

I hope you appreciate the fact that I'm sparing you a long diatribe on the technical aspects of your phone's radio frequency usage. At best, an explanation would be wordy and confusing. And it's unnecessary.

The simple explanation for those letters and numbers associated with your smartphone's signal is this: The higher the number in the radio frequency designation, the faster the data signal can communicate with your phone. The *G* stands for the *generation* of the device. *LTE* stands for Long Term Evolution (basically, industry and marketing jargon), and *4G VoLTE* means that you can make *voice* calls over the LTE data network.

Each leap in technology represents more signal capacity. The faster newer-generation technology also means less *latency* (lag time) between the signal source and your phone. It takes less time to get a signal response, which, in the case of 5G, can mean useful advances in personal medical devices, such as remote monitoring, and more. Also, if you live in a home with lots of people using the Internet simultaneously, the 5G signal has more capacity and will improve buffering (latency) for everyone sharing a single signal.

You don't need to run out and upgrade your phone unless it's so old that the carrier no longer supports it in the country where you reside. *Just FYI:* 2G is gone, and 3G is on its way out, too.

A Word about Privacy and Security

I find it annoying when legitimate publications (online or in print) feature articles that spread fear about using the Internet. Such articles prey on those who aren't tech savvy. Here, I offer solutions on how to keep your phone safe (also, check Chapter 4 for more information).

Free usually isn't

Remember that there's no free lunch in the digital age. Nothing is free — neither websites nor (especially) apps that you load on your phone. For example, those *free* apps can share your information and result in sometimes invasive — and always annoying — advertising. You might feel that viewing ads is "payment" for accessing free sites and apps. That's only partially true: Your usage data is what the company needs and is "buying" from you. Many phone manufacturers, apps, and websites learn about you from tracking your actions on your device.

If you pay for an app or a service on your smartphone — instead of opting for the *free* version — there's *less* chance that your data will be taken advantage of.

The terms of service can be tricky

Every time you start up a new device, introduce a new app, or visit a new website, you're presented with the opportunity to agree to the app's *terms of service* (ToS) and/or privacy policy. You may even see a sentence or two and be prompted to click through to read the entire policy. When did you last read the complete ToS? This document usually spans many pages and is brimming with legalese.

As illustrated in the nearby sidebar, "Nobody reads the terms of service (ToS)," virtually *no one* actually reads the ToS. Most existing policies would take around an hour for the average Android owner to read!

REMEMBER

I strongly suggest that you go over the terms of service and privacy policy when you're presented with them, or at least search the documents to see what they plan to do with your data. The fact that your data may be shared with a government agency or insurance company (now or in the future) is a real possibility.

Being able to confidently assume that every provider on the Internet follows its privacy laws to the letter would be nice, but you know the truth in your heart: There's always a bad egg.

Marsha's sage advice about privacy

Do you want to know the most important pieces of advice I can give you?

» Never share information on the Internet that you wouldn't tell anyone except your closest circle of friends and family — not even your year of birth.

» Don't use free Wi-Fi networks without the protection of a VPN. (See Chapter 16 for more information on VPNs.)

» Beware of texts that want you to click a link to attend to an online transaction or an account. Go directly to the website on your phone's browser, log in, and find out if there really is a problem.

» Do not add free apps to your devices until you're ready to use them. For example, I load the airline's app on my phone only when I'm traveling, and then I delete it when my trip ends. Apps like these can fill up your phone and may drain your battery even when not in use.

Look for more security tips throughout this book!

Chapter **2**

Buying Your Android Smartphone and Accessories

Making a decision on a new smartphone is a big job. You'll probably depend on this little device for the next five years. Because phones are personal tools, when you find the right phone, you don't want to let go. That's okay, too. You can find many uses for older phones — even when you buy a new one.

As you do with a car, you can buy, directly from the manufacturer, a used phone that's "certified preowned." Also like a car, a smartphone loses value the minute you walk out of the store. But a *flagship* (top-of-the-line) phone from a couple of years back may fulfill your every need at a quarter of the new-phone price.

In this chapter, I help you center your search on which phone features and options you need to look for and how to make a solid decision.

Investigate First — Then Make a Buying Decision

You'll find a dizzying assortment of smartphones to choose from in the Android world at varying price points. My goal is to help you narrow the choices, decide where to make the final purchase, and secure the best deal.

Before you decide which make and model you want, take a little time to kick the tires (so to speak) on several choices. I've made the mistake of relying on a (commissioned) salesperson's advice or being tempted by a "good deal" online — only to be less than thrilled when I set up and started using my device. Shopping around and researching the details of this important purchase is your best strategy.

Where do you begin this window shopping experience? The Apple Store is ubiquitous — those folks need to be close to their customers — but can't help you with your Android smartphone search. You may be surprised to know that Samsung Experience Stores exist in some cities; looking there is an option if you happen to have one close. You can try the mobile carrier stores (such as T-Mobile or AT&T), and tell them you're just looking, but you can also visit Costco, Best Buy, Target, Walmart, Sam's Club, or any other big box store to check out its Android smartphone offerings.

REMEMBER

Although the top mobile service providers set up stores everywhere, buying blindly from one of them may not score you the best deal. Retail sales representatives (officially known as *wireless consultants*) may get paid via sales program incentive funds, or *spiffs*, on specific devices and can make up to a 10 percent commission on in-store sales. Being an educated consumer is your best strategy.

Looking at a phone's physical features

When you're not being dazzled by a sales pitch, you can look, touch, tap, and hold phones. (You might be able to play video, too.) Lots of options come into play. Just like Goldilocks, you may need a bit of testing to decide what's "just right" for you. While you experiment with various phones, make notes on your present phone (or on a notepad) so that you can review your thoughts and impressions when you finish your research.

» **Screen size:** I know it sounds silly, but a half- (or even a quarter-) inch of screen size can make a big difference when using a phone. I like a big screen, but I have small hands. (I would prefer a smaller phone, but I have big fingers — good luck on texting efficiently.) Pushing buttons, taking photos, and typing characters can be a completely different experience based on the size of the phone.

» **USB and connections:** If you want to use *wired* headphones or earbuds, you need to plug them in somewhere. Not all phones now have a 3.5mm headphone jack, so you need to use a USB-to-3.5mm adapter. (Chapter 15 has the details on how all this works.) For me, an adapter is just another item I will probably lose.

» **Color:** Gosh — looking at the array of colors on the expensive flagship phones may make you "oooh" and "ahhh." Check out the color assortment in **Figure 2-1**. But you most likely will buy a protective case for your phone (rather than carry it "naked"), so you may never see those colors in their original glory.

» **Display:** Until I wrote this book, I had never realized how much the screen resolution impacts the phone experience. Jumping from phone to phone made clear (no pun intended) which phones were easier to read — and which phones displayed photos better. As with a TV, the conventional wisdom is "The more pixels, the better," but let your eyes be the judge for your screen preference. Write down the phone models that have displays you can see well. Both pixel count and screen type figure into the equation. Why pay for a higher pixel count if you don't have to? Screen technologies can bump up a lower resolution nicely.

Photo courtesy of © Marsha Collier

FIGURE 2-1

Of course, there's debate on the *type* of screen, too. An AMOLED display is considered the best (most expensive) and easiest on the eyes, but an OLED, LCD, or TFT can do the job just fine. If you can, compare different models, side by side, as part of the process. Look at photos (maybe take one or two in the store), read text and news stories on the phone to see which suits your eyes.

» **Camera:** I am not a professional photographer and never need to take a super-high-resolution photo, even for large canvas prints I have at home. A good-quality camera in your phone (with at least a 12-megapixel sensor) is what counts. The latest camera technologies do a good deal of image processing in the background to produce impressive images. High-resolution images take up space in your phone's storage. I try to set the camera for the smallest resolution possible.

» **Storage capacity:** A smartphone has capacity limits. Although your photos can back up to Google Photos (where you can view them over Wi-Fi), you no doubt still want to keep several (hundred?) handy on the phone. Photos, music, apps, videos, and other items all take up space on your phone.

Videos can take up a ton of space on your phone. Apps and music you download also take up precious space. But even with 128 GB, I've never filled up a phone.

» **Expandable storage:** Many Android phones, even budget models, have a tiny slot on the side where you can install an inexpensive microSD memory card. **Figure 2-2** shows the expandable storage card in a tray next to the SIM card.

Expandable storage

Photo courtesy of © Marsha Collier

FIGURE 2-2

For around $15, you can give your phone a 64GB (gigabyte) memory boost! You can add up to 1TB (that's 1 terabyte, or a million megabytes) to the Samsung Galaxy Note 20 Ultra. Even mid and budget phones from OnePlus, Motorola, and more offer expansion of varying sizes. How cool is that? You can have room for lots of videos!

» **Battery capacity:** In general, the bigger the battery, the better. As a baseline, a 4,000 mAh (*milliampere hours*) battery can last you for 15 hours of web browsing and email reading via Wi-Fi without a fresh charge. But, as they say, "Your mileage may vary." Each phone brand can regulate battery usage through the operating system.

How long the battery lasts depends on what you're doing on your phone. Playing music, watching videos, gaming, and having many apps open on your phone make a huge difference in battery usage.

When you add in the type of display and its brightness, calculating the lifetime of a specific battery size usage is tricky business. The website PhoneArena.com regularly updates a web page that tests all phones on their battery test results. You can find the staggering data at phonearena.com/news/PhoneArena-Battery-Test-Results_id124954. The page allows you to compare phones too — a valuable resource.

» **Charging:** You find that some phones offer wireless charging. While that sounds like a great gimmick, wireless charging is slower and less efficient than charging through the power cord. You can read more about charging your phone in the section "Power charging blocks — volts and watts matter."

Lots to think about, eh? I wish I'd known all this stuff when I began buying phones.

Reviewing before making a decision

After you kick all the tires and settle on the features and specifications you want, it's time to find a phone. I recommend taking a bit more time to compare, with the help of these sites, the phones you evaluated:

» **GSMarena,** gsmarena.com: Check this site to see whether an older or less expensive phone includes your must-have features.

» **Phone Arena,** phonearena.com/phones/compare: This site has a good comparison tool and reviews for all phones.

The retail stores I mention earlier in this section all have websites, which are good places to start looking for comparisons and, perhaps, a good deal.

REMEMBER

Reading reviews of phones when the phones are newly released can be misleading. Reviewers are handed certain phone models and then write about what they see — and you *know* it takes a while for idiosyncrasies to show up. Accurate reviews about devices appear only after thoughtful reviewers spend time putting phones through their daily paces.

Considering an older model

I love to get a good deal. Researching topics to write about in this book has taught me a lot more about smartphones than I already knew. Most important is that you don't need to purchase a brand-new phone regularly. I used a 2018 Google Pixel 3 XL running Android 11 as one of my test phones and *really* liked it. See the nearby sidebar "Golden oldies" for a testament to older phones.

Buying a used, older smartphone that's still receiving updates can be a true bargain. I bought my Pixel 3 XL on eBay for under $125. That's a great deal, considering that I got to test out a new (to me) phone brand without having to pay big bucks or sign up for a contract.

REMEMBER

The key to having a device retain full functionality is that it can still accept Android system and security updates. The time frame during which Android accepts updates from major manufacturers and carriers is now reaching four years.

WARNING

If you use an outdated device, enable two-factor authentication on every app you use. Be sure that the older apps you have on the phone can update regularly (or just uninstall them). Consider retiring the phone from phone duty and using it for your music files or just movies.

GOLDEN OLDIES

I do have friends (myself included) who have enjoyed older devices for many years. I use a Wi-Fi 2016 Huawei MediaPad M3 tablet that runs Android 7 daily. A friend, until recently, wouldn't let go of a Samsung Galaxy Note 7, and another, a Samsung Galaxy 6 from 2015. Holding on this long isn't a recommended practice, and the device may not be secure, but when you enjoy using a device, it's often hard to change.

I'm planning to repurpose an old Android phone as a dashcam — and I'm looking forward to seeing the fun videos it takes. All you need is a windshield mount, a car charger adapter, a long cable, and an app like AutoBoy Dash Cam. (The Google Play store lets you know whenever the app you select is incompatible with your handset.). Just be sure not to leave your phone in the car to bake in the sun.

I already converted an old phone into a security camera. Being green (and creative) saves tech from the landfill.

Choosing where to buy your phone

Of course, you can buy new phones from authorized dealers and cell phone providers' stores — just be sure to get all the details when you're offered a bargain. Be sure to verify that the phone you're buying will work with your carrier. (*Note:* Older Verizon and Sprint phones might not work on the current AT&T or T-Mobile network.)

Here are a few off-the-wall ideas for places to find the smartphone you've chosen:

» **Amazon:** You can purchase unlocked, new, or refurbished brand-name phones of all kinds, often at nice discounts. Amazon also sells extended warranties.

» **eBay:** Did you know that phone manufacturers have stores on eBay for refurbished devices? Certified refurbished Samsung phones come with a 2-year warranty and have a 30-day return window! Check out eBay's Brand Outlet stores at ebay.com/b/

Brand-Outlet/bn_7115532402 to see whether other phone brands have also opened a store.

Don't forget individual sellers on eBay, either. Lots of small tech-oriented businesses refurbish and unlock phones. Just search for the phone manufacturer's name and the word *refurbished.* People also sell their perfectly good used phones whenever they buy new ones. Remember that every purchase on eBay comes with a 30-day money-back guarantee. You can add, as an option, a 2-year warranty at a reduced price.

TIP

If you're new to shopping on eBay (or just haven't visited for a while), take the refresher course, of sorts, in my latest book, *eBay For Dummies*, 10th Edition.

» **Samsung Certified Re-Newed:** Ever wonder where all those trade-ins go? The phones are fully refurbished by the manufacturer and sold with a 1-year warranty. You get free 2-day shipping and free returns extended to 15 days after delivery. This site also takes trade-ins, which can lower the prices even more. Visit samsung.com/us/explore/certified-re-newed-phones for more information.

REMEMBER

The best times to buy a new smartphone are when the manufacturers announce new models. They generally offer significant deals on the previous versions. The time frame from Black Friday (the notorious shopping day after Thanksgiving) through Christmas is also when retailers offer deep discounts.

Consider Your Carrier Choice

People tend to stay with the carriers they're used to because they're comfortable with them. But consider this question: Maybe change could be a good thing if you find a cheaper plan or a free phone? In any case, check out all the deals that carriers offer. Also, find out about the carrier's policy on phone locking, as described in the nearby sidebar, "Locked or unlocked phones: What's the difference?"

Checking out the carrier's coverage area

Don't forget to double-check the carrier's coverage maps, by typing in your zip code, to be sure your home has a good connection. Here's where you can find the maps for three major carriers:

» **AT&T**: www.att.com/5g/coverage-map

» **T-Mobile**: www.t-mobile.com/coverage/coverage-map

» **Verizon**: www.verizon.com/coverage-map

Some people say that these maps seem to be less than accurate. For example, I know lots of dead zones in my neighborhood and, according to the maps, I should be just zooming along at high speed.

If you want to find truly accurate information, download the Open-Signal app. Just open the app, and you can see where the coverage is and where the towers are in relation to where you're standing, as shown in **Figure 2-3**.

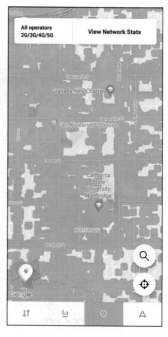

FIGURE 2-3

Finding senior discounts on carrier service

Finding real deals on phones from carriers isn't always easy. Senior discounts are even harder to find. T-Mobile is the only US carrier to have a nationwide *truly* unlimited plan for seniors (currently $70 a month for two lines). As of this writing, both AT&T and Verizon have unlimited plans only for those 55-and-over who live in Florida. You read that right — shall we all pack our bags and move to Florida?

WARNING

Unfortunately, seniors are often a target for questionable marketing, even in the phone carrier business. Brands purchase the right to use the names of famous organizations to make it appear that they are recommended and approved by them. Beware of these practices. I found this disclaimer in the smallest print possible on the bottom of a phone carrier's web page:

"**** member benefits are provided by third parties, not by **** or its affiliates. Providers pay a royalty fee to **** for the use of its intellectual property. These fees are used for the general purposes of ****. Some provider offers are subject to change and may have restrictions. Please contact the provider directly for details."

LOCKED OR UNLOCKED PHONES: WHAT'S THE DIFFERENCE?

While shopping for a smartphone, you may see references to locked and unlocked phones. Previously, whenever you bought a phone from a mobile carrier like AT&T, the carrier would put a software lock on your phone so that you could not switch to a competing carrier, sell your phone, or just give it to a friend. An unlocked phone is a more valuable asset than one that's locked.

Now, customers can legally unlock their carrier-locked phones with no drawbacks. Unlocking doesn't happen by default, but carriers must give customers a way to unlock them. Depending on the carrier, the unlocking process can be easy or annoying. Either way, it can be done — just contact your provider and ask for the unlock procedure. *Note:* The carrier won't unlock your phone until you pay it off.

You Need a Few Accessories, Too

As with most big-ticket purchases, you probably need to buy a few accessories to realize full benefit. For any item that you plan to use over several years, you need to get the right accessories to extend the life of the device.

Power charging blocks — volts and watts matter

Determining whether you have the correct phone charger used to be no big deal. I studied power charging for a few years and found that most of the charging blocks that are supplied with phones delivered the proper charge. This properly matched charger provided, at no extra cost, the exact supply of volts and watts that the phone needed for optimal charging. (The volts and watts of each power block are printed on the sides of the blocks, in miniscule letters.)

Unfortunately, a current trend of manufacturers is to *withhold* the charging block from the purchased phone, as noted in the later sidebar "Who's on (or off) the no-charger bandwagon?"

Behind this no-charger strategy is the assumption that you already have a compatible charger from a previous device. Even if you have an older charger you can use, the reality is that you won't get the benefits of the newest, fast-charging technologies. Yes, you can probably use your existing (lower powered) charging blocks on your new phone — but your phone may not receive the quick-and-complete charging you expect.

The bottom line is that you may have an added expense for a phone charger. Be sure to check out whether your phone comes supplied with the fast-charging option and matching charger. If not, plan to buy one.

WHO'S ON (OR OFF) THE NO-CHARGER BANDWAGON?

The first device to withhold the charging block was the iPhone. An approved, wired, 30-watt charging block for the most recent iPhone costs about $50. Then, if you want a wireless solution, that's at least another $50.

Samsung has jumped on this bandwagon, too. Some of the company's latest phones require you to purchase a charger separately. If you think you can successfully substitute any randomly chosen charger from Amazon, you're mistaken. The Samsung website says: "Using third-party chargers may invalidate your warranty and cause damage." Then you have another $35 expense for a fast-charge block.

At this point, other manufacturers are still supplying power chargers. The OnePlus 9 Pro has a proprietary Warp Charge technology that can get your phone charged from 10 percent to 80 percent in the blink of an eye with the *included* 65-watt charger. Such fast charging seems unbelievable, especially because the OnePlus 9 Pro 4,500 mAh battery is split in two (so you're charging two batteries at once). But it's true. The OnePlus 9 Pro also has a fast wireless charger that will fully charge a phone in 45 minutes. But that charger will cost you an extra $69.

Take the time to establish a phone charging strategy. Lithium-ion batteries don't perform best if you drain them to the teens and then charge them to 100 percent. For the battery to remain long-lasting, charge it after it reaches 20 to 25 percent, and don't allow it to charge past 80 percent. In other words: *Do not leave your phone plugged into the charger overnight,* and, if you plan to store the phone for a lengthy period, be sure it's charged to at least 50 percent, to prevent degradation while turned off.

MicroSD card

If your phone has an external slot for additional storage, buy the fastest and highest-capacity microSD card you can afford. Extra internal storage is a lot more expensive when you pay for having it installed in your phone versus installing an inexpensive brand-name card.

I changed my philosophy on the extra-storage situation. I don't buy a card until I max out my phone's internal storage, and that rarely happens. You can add external storage to your phone at any time. If you take (and store) a lot of videos, installing extra storage might be a good idea. See Chapter 3 for instructions on how to insert a microSD card.

Phone case

As beautiful as your phone is, you might feel safer if you keep it in a protective case. Aside from being protected against impact, your phone may be less slippery and easier to grip when you add a case. If your activities or behaviors are tough on your phone, consider a case graded *Mil-Spec,* which means that it's approved by the Department of Defense for use in the military. And here's one thing I've learned about phone cases: You get what you pay for.

You need to buy a case specifically designed for your phone. If the phone manufacturer makes its own cases, one of these may be your best bet. But you can find other high-quality phone case brands (at a savings) for Android phones. Here are some that stand out:

» **Spigen** makes excellent cases, including a superthin, transparent case that I have used on my phones for years. Spigen also stands by its products. One of my cases cracked on its corner. I emailed Spigen within the warranty period and received a replacement right away.

» **OtterBox** makes incredibly tough and protective cases. My athletic trainer won't keep his phones in any other brand.

» **Speck** makes high-quality cases for a wide variety of phones.

TIP

If you're like me and you want to see the beauty of your phone, you might prefer a clear case. But here's fair warning: Clears cases generally yellow over time and need to be replaced.

Phone Sanitizer

I bought PhoneSoap, the original UV phone sanitizer (see **Figure 2-4**), years ago. I'm not a germaphobe, but I do understand how many germs are on my hands and, therefore, how many must be on my phone. (I've read a widely quoted study that said the average smartphone is 18 times dirtier than a public restroom.)

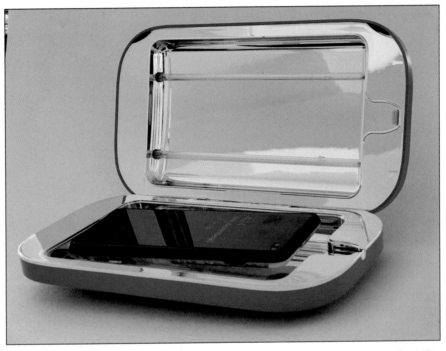

Photo courtesy of © Marsha Collier

FIGURE 2-4

If you wash or sanitize your hands and then pick up a germy phone, you defeat the purpose of a cleansing routine. Using cleansing wipes can be questionable if you miss a spot on the phone, and alcohol can damage the coating of the phone. Especially because many people are now concerned about spreading pandemic germs, a PhoneSoap sanitizer can be an important device to have on hand.

Bathing a device in UV light completely sanitizes it in as little as 5 minutes. A PhoneSoap device kills 99.99 percent of the bacteria and viruses that live on the phone. According to the manufacturer's website, this includes "even the hardest-to-kill germs like E. coli, Salmonella, Staph, Influenza A H1N1, and many more."

I upgraded to a PhoneSoap Pro recently and couldn't be happier. I can also use the device for sanitizing my earbuds, keys, fitness band, and other small items. You can find several PhoneSoap models, starting at $79.95, and you may get lucky and find a discount code on the website at www.phonesoap.com.

IN THIS CHAPTER

» **Unboxing your phone**

» **Inserting a SIM and/or microSD card**

» **Powering up your phone**

» **Connecting to networks**

» **Finding your way around the phone**

Chapter **3**

Activating and Connecting Your Phone

P erhaps few events in life are consistently fun, but opening new (or new-to-you) tech devices can be *exciting*. Manufacturers know this, and the boxes containing new smartphones are designed to be cool-looking. Opening one is a fun occasion.

When I get the chance to open the package for a new phone (or another tech device), I set aside time, grab a cup of coffee (or something stronger), and relish the event. Others (like my husband) can't wait to rip open the package and get their hands on the device.

If you purchase your phone from a carrier (AT&T or T-Mobile, for example), the carrier may offer to set it up for you. But how much can you learn by allowing that to happen? Let them do the basics, then ask questions and then bring it home to complete the setup. If you have a question that's not answered in this chapter, you can always call up someone at the store.

However you've chosen to receive and set up your new smartphone, you may have some questions on your mind. This chapter is intended to answer them.

Unbox Your New Phone

The first step to possessing your new handset is to open the box. Don't just jump to the phone and ignore the rest. See what's included inside the new phone package and make a mental note of all the items you receive with the phone. Then

REMEMBER

» **Pull out your phone** and take in that new design. Peel off the plastic that encases it, and familiarize yourself with the buttons, microphone, and speaker locations.

Some manufacturers install a clear coating (screen protector) over the front glass. It has a hole where the front camera lens is located — don't try to pick at it and peel it off. It's there to protect your phone. If you scratch it, you can easily buy a new one and have a professional install it.

» **Check out the charging system,** if included. Although some brands tend not to offer them, brands that have a proprietary charging system may offer a complimentary cable and power adapter. If one isn't included, be sure to purchase the proper power adapter and cable for your phone (see more about charging your phone in Chapter 2). A flimsy cable and an underpowered charger won't bring joy under any circumstances — and may even be dangerous.

» **Find the SIM-tray ejector pin.** This pin (shown in **Figure 3-1**) is a vital accessory. You need this tool to open the SIM tray to install your carrier's SIM card. (Also, you use the ejector pin to open the tray if you want to install a microSD card with extra storage later on.) In a pinch, you can also use a small, unbent paper clip. Be careful!

SIM card in tray   **SIM tray ejector pin**

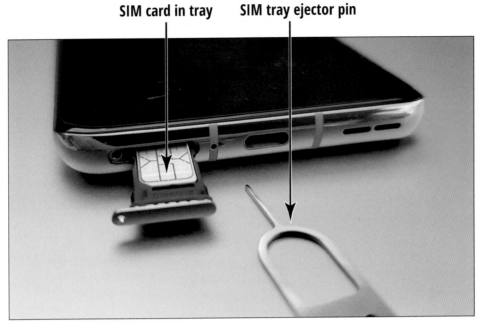

Photo courtesy of © Marsha Collier

FIGURE 3-1

TIP

Keep and carry an ejector pin. I often store one of these tiny tools between my phone and its case — that way, I don't lose it. Should you travel outside the country, you might want to buy a cheap data SIM at your destination and switch it into your phone to save money. (You'll also need it if you install a microSD card).

» **Look for a phone case.** If you're lucky, you also find a free case in the phone package. It isn't usually the case you want to use in the long term, but it can tide you over until you find the one you want.

» **Check for a quick-start guide** (rarely included but nice to have). Usually, you just power up the phone and follow the steps.

Figure 3-2 shows a new phone and its accessories, fresh from the manufacturer's packaging.

Photo courtesy of © Marsha Collier

FIGURE 3-2

TIP

Your phone is precharged when it comes from the factory, so you don't need to worry about charging it. If for some reason the phone lacks sufficient power to complete the setup, you see a warning on the screen and can plug it in.

REMEMBER

When you set up a phone, you can skip steps and set up many options; you have no reason to rush the process.

Insert a SIM Card

If you've decided on a carrier, someone there will have sent you a SIM card to install in your new phone. But you don't need to install a SIM card (if you're still deciding on a carrier), as long as you have a Wi-Fi connection at your location.

Look around the side of your phone and place the SIM tray ejector pin in the tiny hole on the side. The SIM tray pops out, as shown in **Figure 3-3,** so you can insert the SIM card. You find a hole with a notch in one corner which will match your card; it fits properly in only one way. Lay the SIM in place carefully and insert the tray back in the slot the same way you removed it.

Photo courtesy of © Marsha Collier

FIGURE 3-3

Turn On Your Phone for the First Time

As noted in the earlier section "Unbox Your New Phone," you have no worries about your phone being charged (the manufacturer does that for you). Just hold down the power button on the side and watch the phone come alive.

If you bought your phone from a carrier, you probably see an annoying splash screen with the carrier's logo on it. It lasts only a second or two and can be annoying. You can't easily remove it — which is a good reason to consider an unlocked phone.

Your phone begins its start-up routine, and you have to choose a language and answer a question or two. *Hint:* There *is* a difference between English US and English UK — but that option is simple to change in Settings.

Connecting to a network

To continue the initial setup, you must connect to a network of some kind. If your carrier (or you) previously installed the SIM chip, the phone automatically connects with the carrier network. If you have a SIM card and haven't installed it, you can do so now, or you can connect to your home Wi-Fi network. I prefer to set up on Wi-Fi, but either way works.

If you don't put in the SIM, you can click the word SKIP or LATER that may appear in one corner of the screen.

You don't have to perform the entire setup process at one time. Your phone reminds you to complete any requisite tasks after you begin to use it. I learned this because, despite the excitement of firing up a new phone, I'm super cautious about granting permissions and connections.

Some decisions you make during start-up can't be undone easily. As shown in **Figure 3-4**, your phone offers a reminder if a critical element is missing during setup or after. As you can see, you can always tap *No Thanks* at the bottom and proceed.

Setting up a Google account

I believe that having a Google account is at the heart of your Android experience. The account ties in everything you need in order to enjoy all the features of your smartphone. And, you don't have to supply a credit card until you want to make a purchase via Google or Google Play. (You can find my description of a Google account in Chapter 1.)

FIGURE 3-4

Google offers a convenient, safe, and secure way to use a credit card online. As you shop online, Google sees you enter names, address, and your credit card number into various sites and asks your permission to store this information. Then all you have to remember is the credit card code (usually on the back of the card) so that when using it you can verify that you are you.

TECHNICAL STUFF

If you want to enable Google's secure shopping autofill upfront, swipe downward on the phone's Home screen and tap the cog in the upper right corner to go to the main Settings. Then tap General Management ➪ Autofill Service and tap Autofill Service once. You see options for Autofill. Because Google already integrates with your phone, I prefer to use that over the other options you may see. *Note:* If you tap another option, you must agree to new Privacy and Terms of Service. (If you haven't read Chapter 1 on Privacy, now's good time to check that out.) After you select the Google Autofill service, tap the Settings cog next to Google. Then tap Payment Methods and the Add a Card option. Follow the prompts from there. Be sure to enable a screen lock now (see Chapter 4) for security purposes.

Setting up a secure lock

Setting up a secure screen lock on your phone is vital, and you probably have the chance to do so during the phone's setup process. I can't stress the importance enough. Even more than having secure passwords, using a screen lock keeps your phone data private if your phone should go missing.

REMEMBER

Get into the habit of using a screen lock, even if you use your phone mostly at home. You don't want to open yourself up to a security problem if you leave an unlocked phone alone out in the world. Without a screen lock, anyone can get into your phone (and your accounts) just by swiping the screen.

Several screen lock methods are available on most Android phones. I believe they're all safe, but some are more secure to use than others. It's a matter of choice. See Chapter 4 for a list of screen lock options and instructions for enabling them.

Restoring data from an older phone

During setup, your phone asks whether you want to restore the data from an older phone that has been backed up and registered on your Google account. You see a screen showing older devices you've owned (see **Figure 3-5**), and you can select one to restore from.

TIP

Consider each brand-new phone a clean slate — I prefer to do so, so the only info I restore from an older phone are my Contacts. After you use a phone for a while, you may collect extraneous apps and messages that you don't need. (Your email follows through on the Gmail app.) Plus, unnecessary data takes up room on your device. The more space you give the operating system, the faster the phone responds.

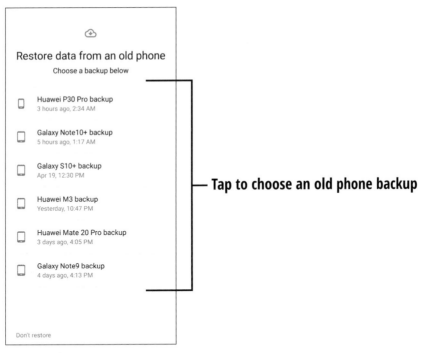

Restore data from an old phone

Choose a backup below

Huawei P30 Pro backup
3 hours ago, 2:34 AM

Galaxy Note10+ backup
5 hours ago, 1:17 AM

Galaxy S10+ backup
Apr 19, 12:30 PM   — Tap to choose an old phone backup

Huawei M3 backup
Yesterday, 10:47 PM

Huawei Mate 20 Pro backup
3 days ago, 4:05 PM

Galaxy Note9 backup
4 days ago, 4:13 PM

Don't restore

FIGURE 3-5

Learn Android Smartphone Symbols and Gestures

Some basic commonalities among Android smartphones just don't change (thank goodness). Once you know the commonalities, you recognize them from version to version — which makes updating your phone ever so much easier. I'm sparing you the official nerd jargon because, frankly, we don't need to worry about it. What we need to know is how the commonalities work and which actions they perform.

Recognizing common Android icons

Table 3-1 shows you some common symbols (*icons*) that you find littered throughout Android devices.

TABLE 3-1   Common Android Symbols

Symbol	What It Does
▶	Plays video
🔊	Turns volume up or down or turns sound on
🔇	Turns sound off
✎	Opens editing options
≡	Sorts items on a page. alphabetically or by another spec
≡	Offers more options when tapped. (This *Hamburger* icon may also appear as three dots stacked vertically.)
🗑	Deletes a selected item
🎤	Listens for voice dictation, search terms, and other sounds
✕	Closes the screen where it appears in the upper corner
★	Indicates a Favorited item
🔍	Searches a screen or an app
✓	Dismisses a screen, indicating that you have completed an action
⚙	The infamous Settings Cog icon; opens settings for the system or for individual apps
⬟	Shares content to messages, email, social networking, apps, printing, and more

Meeting the top status bar

At the tippy-top of your phone, you find a bar with tiny icons. **Figure 3-6** shows examples of these bars. The icons are tiny and, frankly, can be hard to see. A few convey some valuable information.

» **Time display:** The time of day as sent by the carrier

» **Wi-Fi signal:** If you have Wi-Fi on, you see an upside-down V with bars to indicate the strength

» **Carrier signal bars:** A triangle that grows taller (with bars) as the carrier's signal grows stronger

» **Battery:** A small battery symbol that becomes lighter or darker (depending on the background) as the battery loses juice; or, on certain phone models, also shows the percentage of power remaining

» **Network speed:** Shows how fast (or slow) the network connection is; not available on all devices

» **Voice over Wi-Fi:** If your Internet signal is stronger than your cellular signal, some phones will take advantage of that and place the call over your Wi-Fi. This capability establishes more reliable calling if your carrier permits this type of connection.

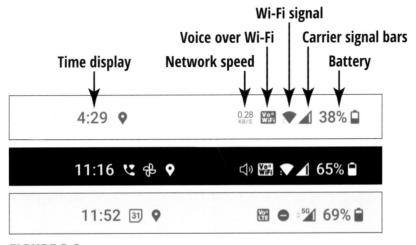

FIGURE 3-6

As you can see in this book, Android has a powerful Settings manager. Some brands of Android phones even let you add icons to this top status bar.

Exploiting screen navigation and gestures

As part of starting up your phone, you are required (this is *not* an option) to decide between using gestures (swiping your hand across the phone) or visual buttons to navigate.

I'm all for hand gestures, but because they can be used for other activations on my phone, I prefer also using the navigation buttons at the bottom of the screen. For example, a hand swipe can be used (on some phones) to take a *screen shot* (an image of the phone's current screen) for recordkeeping purposes or to share with someone. You can always take a screen shot by pressing the Volume Down and Power buttons as the same time.

TECHNICAL
STUFF

If you have Google Assistant installed, you can always say, "Hey, Google. Take a screen shot." Find information about Google Assistant in Chapter 7.

Setting up your choices for taking screen shots varies between phone brands. Swipe down on the Home screen, tap the System Settings cog, and follow the appropriate actions for your phone from there:

>> **Samsung:** Settings ⇨ Advanced Features ⇨ Motions and Gestures ⇨ Palm Swipe to Capture

>> **OnePlus:** Settings ⇨ Buttons & Gestures ⇨ Quick Gestures ⇨ Three-Finger Screenshot. This setting enables you to run three fingers from the top of the phone to the bottom to take a screen shot.

>> **Xiaomi:** Just run three fingers from the top to the bottom of the screen.

Notice that you can also set many other options for gestures. You can explore these options by searching for the word *Gestures* on the main Settings screen. **Figure 3-7** shows the selection screen.

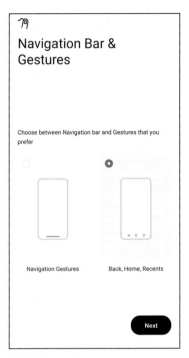

FIGURE 3-7

The bottom-of-screen navigation option offers these three easy to see buttons (shown in **Figure 3-8** as they appear on two phones):

» **Back:** Tap the Back arrow (triangle) to navigate to the previous screen you were viewing.

» **Home:** Tap a circle (or square) in the middle of the buttons to return to the Home screen.

» **Open apps:** Tap three bars or the square icon to see all open apps. You can then tap to use one or close them to save memory space.

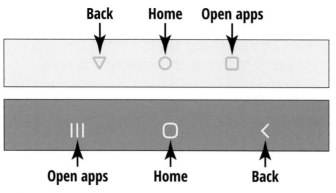

Back **Home** **Open apps**

Open apps **Home** **Back**

FIGURE 3-8

2

Getting Started with Your Android Smartphone

IN THIS PART . . .

Putting personal and data security first

Setting up screen locks

Discovering your phone's buttons and apps

Managing the Home screen

Choosing and using a favorite keyboard

Engaging Quick Settings and Google Assistant

IN THIS CHAPTER

» **Employing your phone's security features**

» **Choosing a screen lock type**

» **Backing up data**

» **Trying out sophisticated biometric security**

» **Adding owner and emergency info**

Chapter **4**

Safety First: Making Your New Phone Private

As you flip through this book, you might notice that I am quite security oriented. I believe in preventing security problems before they occur. Too many times, you can avoid a security problem, such as hacking, by taking simple steps at the start of your smartphone journey — rather than after an unfortunate incident occurs.

Remember that the marketing spiel behind a phone's built-in security features often outweighs the reality. Not every feature is failsafe, but you need to make the most of them, considering the importance of the contact information and other data that you keep on your phone.

In this chapter, I help you establish real-world practices that can engage or supplement your phone's safety features and thereby protect your private data. No need to leave an open door to your privacy.

Set Up a Screen Lock

When you start up a brand-new smartphone, one step in the process is to set up the *lock screen*, which secures the device against unwanted users. The lock for the screen can take various forms, but in each case, you determine the key, of sorts, that's needed to unlock it. You may need to use this key when you wake up your phone, authorize purchases, or sign in to certain apps, for example. Some people skip this lock screen step, perhaps thinking, "I'll do it later" — or even worse, "I'm mostly going to use my phone at home, so do I need to worry about someone else turning it on and accessing my bank account?"

My answer to this question is that you need your phone to be locked — *always.* The newest locking tools are far less complex than ones in the past and are far more secure.

Checking out the screen locking options

You have several options for a screen lock type, and my research shows that many phone brands support all types in this list:

» **Swipe:** This way, you can unlock your phone by simply swiping the lock screen upward and opening your phone. This is zero-level security, and anyone can swipe your phone and then access your data.

TIP

So that you have alternatives for accessing your phone, you can set up more than one type of screen lock.

» **Pattern:** Just trace a line with your finger from dot to dot to dot. The catch is that you need to remember the pattern every time you want to unlock the phone — even when you wake up from a deep sleep. I've never used this one, because I'm sure I'd forget the pattern. Maybe your memory is better than mine.

» **PIN (personal identification number):** What you must remember when using this *mostly* secure method of securing your device is a series of numbers. How many numbers is usually up to you, but the minimum is four. Don't use your street address or birthday. (Maybe pick your great-grandmother's birthday instead – at least that probably doesn't appear on the Internet.)

» **Password:** You can use a password, but a simple one isn't secure. You can attain a high level of security with a complex password, but who really wants to type a complex password every time they open their phone? Not me.

» **Biometric fingerprint:** Face it: You've probably given up your fingerprints before, somewhere. But although you may have given them up, fingerprints are nearly impossible to replicate, making fingerprints the most secure method of locking your phone. According to *Scientific American*, there's a 1-in-64-trillion chance of a duplicate fingerprint existing. It's interesting that even though glass fingerprint sensors are common, Apple's newest iPhones have excluded this option in favor of facial ID. **Figure 4-1** illustrates setting up locks for the PIN, biometric fingerprint, and pattern lock types.

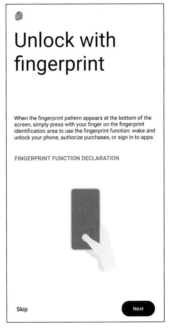

FIGURE 4-1

REMEMBER

You should register more than one fingerprint if you use the biometric fingerprint lock type — accidents do happen. I once lost a fingerprint for a while after touching the stove and burning my finger.

» **Face recognition:** In the United States, most people aren't accustomed to having their faces scanned and recorded. And as a matter of privacy, many are uncomfortable with the idea of this activity. Face recognition on a smartphone is done via software versus biometric identification. It's trendy, but your phone can be fooled and then unlocked by someone holding a picture or video of you to your phone's camera.

Be aware that facial recognition on a smartphone is not recommended as the most secure lock option. According to Samsung's website support, "Face recognition is less secure than Pattern, PIN, Iris, or Fingerprint." Of course, the screen lock type is *your* choice.

REMEMBER

According to Apple, "Face ID requires that the TrueDepth camera sees your face, whether your device is lying on a surface or you're holding it in a natural position." This indicates that the phone's camera must be on when you're unlocking the phone. I'm sure the Android version works in a similar fashion. I only want my camera on when I want it on.

Many folks I know set up fingerprint *and* PIN-number screen lock types. These are both secure choices and easy to use.

Following the lock screen setup process

While you're setting up a new phone, you walk through the lock screen security configuration. If you didn't set up screen locks during start-up, here's how you can do it now:

1. Swipe downward on the Home screen and tap the main Settings cog to open your phone's general settings.

2. Tap Security & Lock Screen (or Biometrics and security), and then tap the appropriate option for setting a lock type.

If you've previously set up any form of screen lock, you now must enter the PIN, pattern, or password before proceeding.

3. On the resulting screen, tap the screen lock option you want to use and follow the onscreen directions.

If you want to set up the fingerprint screen lock type, that's a little more complicated. You need to register the entire print from a screen, as shown in **Figure 4-2.** If you chose the fingerprint option, follow these steps:

1. Place your finger over the fingerprint icon on the screen.

2. As you press down on the glass, you can watch your fingerprint fill in on a graphic above it.

3. When that particular segment of your fingerprint is recorded, you feel a vibration, indicating that you should move your finger a little to record more print info.

FIGURE 4-2

4. Repeat the finger press as you see your print fill in on the screen. When your phone has recorded the print, it lets you know.

Now your phone is secured, and unless someone knows your PIN or can replicate your fingerprint (à la James Bond), they can't unlock your phone and access your data.

Establish Data Backup

Backing up your phone's data can be important to retaining contacts, photos, and other information. Again, you have a choice to make. If you plan to forevermore purchase phones from your current man-ufacturer, by all means, back up to the brand's cloud service if they have one. (You're most likely prompted to do this.)

If you think you might switch to another manufacturer's phone (even an iPhone), you might be better off using Google Backup. The space taken up by Android phone backups isn't counted toward your Google Drive space allocation (which is also free when you set up an account).

Generally, the backup to Google is automatic, but if you want to double-check, here's how:

1. Swipe downward on the Home screen and tap the main Settings cog.

2. In the open Settings app, scroll to find the options named System or Accounts and Backup. Tap the entry you find. (You can also just search for *backup*.)

3. Tap Backup and you see a screen similar to the one shown in **Figure 4-3**. This screen may already show your Google backup infor-mation. If it doesn't, find and activate the toggle next to the backup option.

4. Make sure that your Google account is listed, and then you're ready to roll.

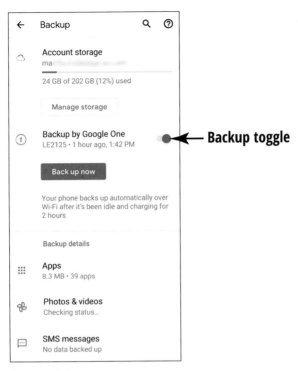

Backup toggle

FIGURE 4-3

REMEMBER

Apps such as Google Photos continually back up on their own to the Google Photos cloud (a portion of Google Drive).

From this point on, your phone seamlessly backs up your contacts, Google Calendar events and reminders, Wi-Fi network information, app display settings, and more. The backup comes in handy if you ever need to retrieve this information — or whenever you buy a new phone. On a new device, you can decide which information you want to carry over from your old phone's backup.

Place Owner Information On the Lock Screen

In this section, I describe a little-known tool that is my preferred version of emergency contact information. In case you lose your phone, the safest way to get your phone back is to make your owner information is visible on the screen without someone having to unlock your phone first.

I add this owner information to all my phone screens — and I just hope it makes a difference. If I ever leave my phone somewhere, I'm hoping that a kind soul picks it up, sees the notice on the lock screen, and contacts me.

As shown in **Figure 4-4**, a message on my lock screen reads: "REWARD: If found call 310-***-****." You may also choose to add an emergency contact's name and info to the message.

 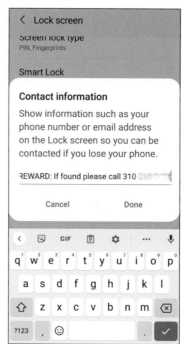

Samsung Galaxy S10+

FIGURE 4-4

Here's how to set up owner information on the lock screen:

1. Swipe downward on the Home screen and tap the main Settings cog near the upper right corner.

You can also find Settings by tapping the Settings app icon on the App Drawer screen(s) you access by swiping upward on the Home screen.

2. At the top of the resulting screen, tap the Magnifying Glass icon (indicating a search) and type *lock* in the text box next to it.

Because various brands offer varied ways to wade through the main settings, the most convenient way to search for a specific setting is to use the search box you find to the left of the magnifying glass at the top of the main Settings screen.

3. On the results screen, scroll to find an option named Add Text on Lock Screen (or Contact information). On Samsung phones, tap Lock Screen to bring up a controls screen (see **Figure 4-5**), and then tap Contact Information.

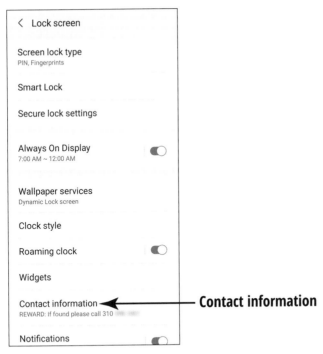

FIGURE 4-5

4. In the text input area you see, type a short owner-contact message. Tap Done when finished with your message.

This contact message appears whenever someone touches the lock screen on your phone. I believe that the information in this message can be the best chance at getting your phone back — or of being identified in an emergency situation.

TECHNICAL STUFF

The phone number I use as my owner information is a Google Voice number. Any device I own (even an Apple device) that has the Google Voice app receives, via email, a transcription of any voicemail message sent to that number. You can also set Google Voice to forward calls to ring numbers you have previously linked in your account. Chapter 1 gives you details on how to secure a free Google Voice number.

Add Emergency Info

While researching this portion of the book, my goal is to be sure that every piece of information I give you is actionable. I believe (and I read everywhere on the Internet, from reliable sources) that setting up an ICE (*in case of emergency*) contact on my phone is important. The newest phones have places to input your medical information for use in case of an emergency.

Designating Emergency or ICE contacts

For safety's sake, go to the Contacts app and find your family's contacts. Tap a contact's name and, at the very least, tap the Star icon to designate that person as a Favorite. All Android phones offer this choice. If you see an option for Family or Emergency, tap that option as well. Your favorites should appear at the top of the Contacts app list.

You can also indicate emergency contacts from within the Contacts app on Samsung phones (it's the orange Contacts icon) by tapping the left side Hamburger menu (three lines) and selecting Emergency Contacts from the Groups menu. You can then add them to appear at the top of the home screen in the Contacts app.

When you tap Emergency Contacts from the fly-out menu, you also have the opportunity to fill out the full emergency info. In the Google Contacts app, the flyout menu reflects labels. You can add labels for groups like Family or ICE if they are not already there.

TIP

I use the (blue icon) Google Contacts app — which you should too because then all your contacts port to all the Google apps that you use. I lost a bunch of contacts once by accidentally putting them in my phone's contact app, and they didn't appear when I moved to a new phone from another manufacturer.

Providing medical information

The odds of someone — specifically, emergency professionals — spotting emergency information on your phone are not high. I outline my observations from primary research in the nearby sidebar, "What the first responders say."

WHAT THE FIRST RESPONDERS SAY

I went to the sources who would most likely need your emergency information — and found out that

- *Neither emergency medical technicians (EMTs) nor paramedics typically search for info on your phone.* Their job is to treat you and stabilize you enough to reach the hospital. These professionals are way too busy to poke around on your phone. I spoke to many first responders at fire stations and asked, "Do you look at the patient's phone for emergency information?" The answer was generally "Heck, no."

- **If you're in a car accident, the police may identify you by "running" your license plate through their verification system.**

If you're conscious in an emergency, you can offer the PIN lock code on your phone to doctors and nurses in the hospital so that they can find information on your device. I keep my medical info in my Google Keep app (and a paper backup folded up behind my Driver's license in my purse).

REMEMBER

The safest method of supplying emergency medical information is to keep it in your wallet, next to your identification cards. Also, paramedics *may* look for a medical bracelet.

To fill out the emergency information (and contacts) on your phone, follow the steps after the figure. Remember that you can fill in all or only part of the form, an example of which is shown in **Figure 4-6**.

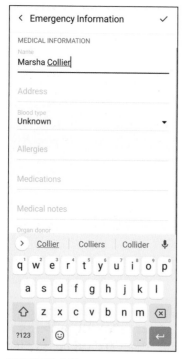

FIGURE 4-6

On most Android phones, do this:

1. Swipe downward on the Home screen and tap the Settings cog near the upper right corner.

2. After the Settings app opens, tap the Magnifying Glass icon and type *emergency* in the Search text box. The option to fill in emergency information (and emergency contacts — same form, refer to Figure 4-6) appears.

3. Type as much or as little information as you want to reveal. When finished, confirm that you want to add the information when prompted (see **Figure 4-7**).

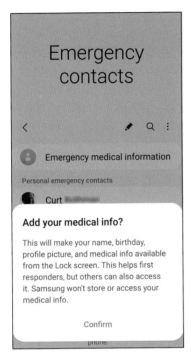

FIGURE 4-7

WARNING

If you fill out the emergency information, anyone who picks up your phone can access this data (your name, birthdate, and medical info). Your private data in this area is unprotected from prying eyes.

Chapter **5**

Personalizing Your Handset

Before delving into using the nuts-and-bolts (or, the functions and features) of your phone, getting to know the hardware buttons and software apps that come preinstalled on your phone is important.

Your phone may not be brand-new, and you've no doubt noticed many of its icons, controls, or buttons. But you may be at a loss to determine what they all do. To be honest, I often discover new tools on my phone. These can appear whenever incremental (and major) updates from Android add features to make the phone more useable.

After I got the hang of using the basics of the Android operating system, I found that newly installed features no longer throw me for a loop.

In this chapter, I start out with the essentials. When a feature needs further explanation, I let you know in the text which chapter to go to for deeper exposition.

Hardware Buttons and What They Do

Explaining a button's location and function is a bit tricky because Android is an *operating system* (software that supports a device's basic functions) and not a phone manufacturer. Each manufacturer likes to put its own spin on design and ergonomic operation of the phone. Going from one brand to another can force changes in your long-term habits. Although annoying, I look at this change-up as a good thing; learning new ways to do things helps keep my brain sharp.

As long as you know what to look for, a little poking around can help decipher where the buttons reside on each design. **Figure 5-1** shows the layout of the power and volume buttons on a group of smartphones. *Note*: Pressing certain buttons in combinations have functions, too. For example, pressing the power button on the right and the volume down button at the same time can trigger a screenshot. If you take a screenshot accidentally, you can easily delete the unwanted one in your Gallery app. (Find out more about the Gallery app in Chapter 12.)

TIP

I'm a big fan of the *Mute switch* — a hardware button that is exclusive to the OnePlus brand (and is similar on the Apple iPhone). The Mute switch is a silver switch on the right side of the phone, as shown in **Figure 5-2**, and by toggling the switch, you can set the phone to mute or vibrate instantaneously. I often find myself scrambling to silence my phone when it goes off during a meeting, a ceremony — during a plethora of embarrassing times. Best of all, I can use the Mute switch to silence my phone in a movie theater without lighting it up and working my way through a menu.

Power/Bixby button

Volume rocker

Bixby button   Power button

Volume rocker

Power button

Mute button

Volume rocker

Samsung Galaxy Note10+   *Samsung Galaxy S10+*   *OnePlus 9 Pro*

FIGURE 5-1

OnePlus 9 Pro Photo courtesy of OnePlus

FIGURE 5-2

Power buttons

The power button appears as a single button on the right or left side of the phone (refer to Figure 5-1). Holding down this button usually opens a screen listing three (or maybe four) options. **Figure 5-3** shows two 3-option examples. The options function as described in this list:

» **Power off:** Simple — a long-press on the words *Power Off* will bring you to the power off screen on your phone. Despite the tech myths, powering off your phone at night doesn't prolong the battery life much. There's really no need to power off your phone unless you're swapping SIM chips or performing physical repairs.

TECHNICAL STUFF

SAMSUNG GALAXY BIXBY AND HOW TO CHANGE THE BUTTON

On Samsung devices you can find a hardware button in the middle of the left side of the phone. This button will activate Samsung *Bixby Assistant*, the Samsung version of Google Assistant. Bixby has a following and is now featured on Samsung's newest smartwatches. The new range of Samsung smartphones have no dedicated button.

If you've never set up Bixby or you're using an older phone with the button, you might not want to bother with it. Remember that when you set up the Bixby Assistant, Samsung requires you to also set up a Samsung account (with a new set of privacy settings). The Google apps do a good job for me, so I see no need to venture into another account — again, my preference is not to share any more personal data than absolutely necessary.

This *is* Google's Android phone. If you want the Bixby key to do something other than trigger Bixby, such as open the camera or another program or even power off, you can remap it without setting up Bixby.

To remap the Bixby key on certain Galaxy phones, follow these steps:

1. Swipe down the Home screen to reveal the Quick Settings.
2. Tap the Advanced Features section to open the Side Key Settings menu.

3. Tapping the Side Key Settings opens a menu where you can change the button to power off. You can also assign a double-tap on that button to open an app or launch the camera.

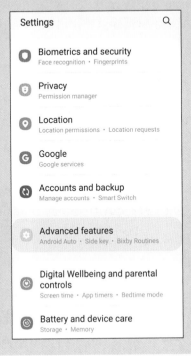

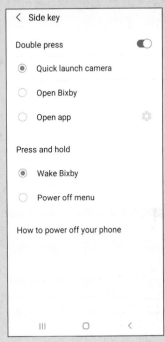

REMEMBER

Don't reach for your glasses if your phone does not match some of the screenshots in this book. These screens were captured from phones running Android 11. If you have an older (or newer) model phone, your screens may be similar, but not exact.

» **Restart:** Your phone may start acting finicky and you need it up and running at full speed. Restarting your phone will do the trick and clear things up.

TIP

Restart your smartphone at least once a week. Consider it a refreshing cleanse of your phone. When you restart, it clears out old memory — remnants of programs and applications that may be draining your battery. Restarting also helps your phone function better, especially if you notice that it's running slowly or acting strangely.

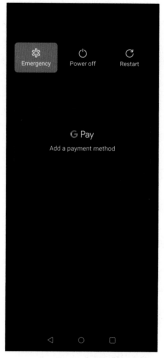

Samsung Note 10+   *OnePlus 9 Pro*

FIGURE 5-3

» **Emergency mode:** On the flagship phone models, which are the newest and most expensive, you may see this option. Tapping Emergency Mode opens a screen that offers these options:

- *Add your emergency medical information.* If you've read Chapter 4, you know why I recommend not doing this.

- *Other emergency settings* are a good idea to set up. You may have options that would help you in a rescue situation: a flashing SOS light and/or an SOS sound, for example. You may also be able to input contacts to receive an automatic text message with your location information in case of an emergency.

» **G Pay:** Some Android smartphones, such as OnePlus and Pixel (made by Google), may have the option to set up a secure payment method for Google Pay (not to be confused with Google Play store) on this screen. (Samsung doesn't offer this option because it has its own payment platform.)

You can use G Pay for online purchases or contactless purchases from your phone. Or you can choose not to set it up just now. In the future, G Pay plans to offer electronic boarding passes, IDs, event tickets, and more. Similarly to the iPhone's Apple Pay, G Pay lets devices communicate with retail point-of-sale systems, too.

You may see variations on this screen as you receive new phone updates, but the basics I mention here will be there for the near future.

To get to the screen to power off a Samsung Galaxy Note 10+, hold down the Lower Volume key and the side (Bixby) key at the same time until the screen appears.

Volume buttons

You can easily find the Volume button(s) on your phone (refer to Figure 5-1). They generally live toward the top of the phone, on either the right or left side of the handset. I've seen volume controls as two separate buttons or as a single rocker button that turns the volume either up or down.

Hold down the top button (or the top of the rocker button) to make sounds louder, and hold down the bottom button (or the bottom of the rocker button) to lower the volume.

The change in volume usually applies to your current task — for example, listening to music or quieting your ringtone. When used during a task, the volume alteration affects only the function you're using. You have this single button (or set of buttons) to control the volume for (commonly) these four distinct functions:

» **Ringtone:** A sound that accompanies an incoming call. (How loudly do you want your phone to ring?)

» **Notification:** The sound you hear for a text message or another alert that your phone receives

>> **System or alarm:** A sound your phone might make whenever you touch the phone or tap a key

>> **Media:** Movies, games, and music, for example

Never fear! You can adjust more than one function at the same time from the main touchscreen or from within the main phone settings. On the touchscreen, tap what you may see as a Settings cog icon or a set of three dots that appear in the volume control when you hold its button up or down. After you tap the cog or the dots, options appear. These options can vary depending from where you access them, as shown in **Figure 5-4.**

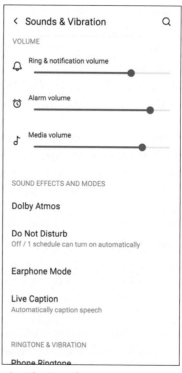

Samsung Note 10+ from Volume Control   OnePlus 9 Pro from main Settings cog

FIGURE 5-4

You can also customize the volume and sounds on your phone in the main settings area, which you access by tapping the Settings cog in the pull-down shade (the Quick Settings area that rolls down when you slide your finger twice from the top of the phone). You can find more info on using the tools in Chapter 8.

Find and Sort Your Apps

Though iPhone owners have their apps on the Home screen (you can swipe left through multiple screens to see all installed apps), most Android phones use the Apps drawer. The Apps drawer is where you can access all your apps, whether they're preinstalled or installed by you. You can duplicate app icons and move the resulting shortcut icons to the Home screens (see the later section "Moving App Shortcuts to the Home Screen"). The Home screen generally has three or four screens that you scroll through by swiping right or left. These screens can hold shortcuts to your selected apps.

The app icons on the Home screen(s) become just links to the apps that live in the Apps drawer. You can delete the app icons from the Home screen, but that action doesn't uninstall the app. To uninstall an app, you must go to the Apps drawer. I tell you more about uninstalling apps in Chapter 16.

Depending on the brand of phone you're using, you may see the symbol shown in **Figure 5-5** at the bottom of the Home screen. Tapping this Apps icon brings you to the area where the phone apps reside.

 ← Apps icon

FIGURE 5-5

If you're missing this icon, try swiping the screen from bottom to top. That should do the trick. You should see a screen like the one shown in **Figure 5-6**. As you add apps to your phone from Google Play store, they show up there.

FIGURE 5-6

Samsung phones have a useful feature that allows you to customize how apps are sorted. Follow these steps:

1. Swipe up from the bottom to the top of the Home screen, and then you see the Apps drawer.

2. In the upper-right corner of the resulting screen, you see three vertical dots.

3. Tap the dots to reveal a menu with several options, which vary depending on your phone model. Some Android phones offer you a quick link to the main phone Settings, along with a Sort option. (Refer to Figure 5-6.)

4. Tap the Sort option and you're given a choice to personally customize the sort or have the apps appear in alphabetical order. (I am a fan of alphabetical order.)

Move App Shortcuts to the Home Screen

Because I use social media regularly, I want to be able to quickly access the related apps I need. Instead of my having to go to the Apps drawer, I want these apps (such as Twitter) available on my phone's Home screen. To get them there, I need to make shortcuts. The process to do this is quite simple:

1. Swipe up (or sideways depending on your device) to open the Apps Drawer screen, which shows your apps. (It's not really a drawer, is it?)

2. Find the app you want to place on the Home screen by scrolling through the Apps Drawer screens. (The scroll may be vertically or horizontally, depending on your phone.)

3. Place your finger on the app icon and *long-press* (hold down your finger). You see a tiny pop-up menu appear, but for now just ignore it. Your phone creates a duplicate moveable icon that becomes the shortcut.

4. Slide your finger (while still pressing down on the app icon) to the edge of the screen, and you should arrive at the Home screen. Lift your finger, and the app shortcut has a new home. At any time, you can change the position of any app icon — even on the same screen — by long-pressing the icon and moving it with your finger.

   If there's no room on one Home screen, your icon will be directed to another. You can swap out shortcut icons — from screen to screen — easily.

 Some Samsung devices enable you to add apps to the Home screen by long pressing on the chosen app's icon and selecting the relevant option from the menu that appears. As shown in **Figure 5-7**, the option is Add to Home.

FIGURE 5-7

If you move an app unintentionally to the Home screen, no worries! Just long-press on the app's icon, and a new pop-up menu appears, as shown in **Figure 5-8**. Tap Remove, and the shortcut icon disappears from the Home screen. The main app icon still appears in the Apps Drawer (er, on the apps screens).

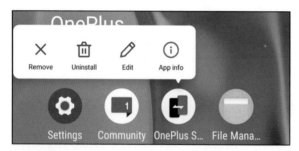

FIGURE 5-8

Group Apps into Folders

Although you can group many apps into folders on the Home screen for organizational purposes (the same as on an iPhone), I tend to use this option sparingly. I use it mostly for quick access to apps I use almost every day — the same apps that I add as shortcuts to the Home screen. You may notice that your phone came with some app folders already installed. These folders, as shown along the top in **Figure 5-9**, can appear on the Home screen or may be preinstalled at the top of the first page of the Apps Drawer screens.

FIGURE 5-9

You may decide that you want to create your own app folders because, after a while, the Home screen can become a bit crowded. Start building a folder by bringing a couple of app shortcuts to the Home screen as outlined in the earlier section "Move App Shortcuts to the Home Screen."

The example in this step list shows the folder I created for easy access to the social media shortcuts (Instagram, Facebook, and others) that I moved to my Home screen. To try it yourself, follow these steps:

1. Long-press on an app icon that you want to place in the folder with another.

2. Dra-a-g that icon on top of the other app icon, and instantly you see them placed together into a folder.

3. When the folder opens up, type a name for the folder. **Figure 5-10** shows that I named my folder Social Media.

FIGURE 5-10

4. After you name the folder, you can add other app shortcuts if you
   - Tap the plus sign (+) and then tap an icon from the screen that appears. Tap Done and the icon is added to your folder.

- Dra-a-ag an icon from the Apps Drawer screen and drop it on the folder.

5. When you're done adding apps to the folder, tap anywhere on the screen to close the folder.

The app folder holds teeny-tiny icons of the apps you've stored in it. **Figure 5-11** shows folders with teeny-tiny icons.

Google   Microsoft   Samsung   T-Mobile

FIGURE 5-11

Deal with Preinstalled Applications (or Bloatware)

When you buy a new smartphone, the manufacturer — or the carrier — may preinstall popular apps on it. Some of the apps may also be specific to the manufacturer or to the carrier you use. You may or may not want to use these proprietary apps.

Here's a good example: Some phone manufacturers include their own version of the Calendar app in competition (for your data) with Google Calendar. Differentiating the two apps can be tricky, so I like to hide or disable the one I don't want to use.

Whether you want to use the bloatware apps or not, they're innocuous but rudely eat up space and drain resources on your phone. After a couple of years, you might prefer to have that space for your own use.

What to do? Many of these preinstalled apps cannot be uninstalled, so you have to disable them (although some manufacturers prevent you from doing so). To disable an app, follow these steps:

1. Press and hold the app icon until a pop-up menu appears.

2. Tap Select (or Select items) from that menu, and the screen changes, allowing you to select individual apps by tapping on the small circle in the corner of an icon.

3. Select the offending app(s), and click Disable if that option appears at the top of the Select screen. If the word *Disable* is grayed-out (or does not appear), long tap the icon and drag to create a new folder with other unwanted apps.

If you can't uninstall an app, be sure you don't have a shortcut on the Home screen. (If an app was preinstalled, just delete its icon by selecting it — touch and hold to bring up the options — and tapping Remove.) Then create a folder named Apps I Don't Like on one of the Apps Drawer screens, and you can quietly nest them away.

Get the News (and Other Media) You Can Use

Swiping from left to right on the Home screen brings you, in most instances, to the Google Discover news feed. This feed is populated with top stories from various news sources. As you read articles in any of the news feeds on your phone, Android (Google) gets to know which articles interest you, by using artificial intelligence, or AI, to discern your preferences. The feed then aggregates news to your interests.

TIP

You can download the Google News app from the Google Play store, which then appears in an apps screen. (You can duplicate the shortcut icon on the Home screen, as I do.) The Google News app gives a deeper dive into the news. You can not only see top stories but also select headlines on an assortment of topics and more.

SAMSUNG FREE: NEWS, TV, AND GAMES

On some newer Samsung phones (or ones running Android 11 and the ONE UI 3 update), swiping right from the Home screen reveals a new feature from Samsung: Samsung Free, shown on the left in the sidebar figure. This feature offers access to TV shows from Samsung TV Plus, news articles, and games.

When you first swipe to open Samsung Free, you see a trailer and a link to a robust privacy notice indicating how they may share your information. If you're okay after reading the terms, go back and tap to check your agreement with each point on the main page. After you click Agree, you gain access to all these services.

Look for the three tabs labeled Watch, Read, and Play. They allow you to customize the type of content you want to (you guessed it!) watch, read, or play.

If you want to remove this feature from your phone, tap on any blank space on your Home screen, then swipe right and you'll see that Samsung Free has an on/off toggle (as shown on the right in the sidebar figure). Toggle it off, and it's gone. The Galaxy S21 offers an option to replace it with Google Discover.

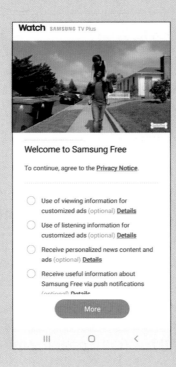

Customize the Home Screen with Widgets

A brand-new phone can leave you with a barren Home screen. You can doll up the Home screen so that you have not only shortcuts to apps you want but also on-the-spot notices and links. This is part of the magic of Android.

First thing out of the box, I add these widgets to any phone I have:

TIP

» **Time:** This one gives you a clock so that you always know the time. Often, you can choose from many variations of clocks, so pick one that makes you smile — or just does the job.

When you travel, certain time widgets show you a clock with the current time in your home city and — on another clock — the time of the city you're in. How handy! **Figure 5-12** shows an example.

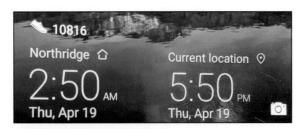

FIGURE 5-12

» **Weather:** You need to know the weather, right? I do — at all times! By having a weather widget on the Home screen, you can see your hometown weather at a glance. Tap it, and a screen with more details appears. You can even save other cities so that you can keep track of the weather where your cousin lives.

» **Google Search:** Even though I use Google Assistant for searches by speaking my search (I show you how in Chapter 7), I often want to type in a quick search without opening my browser.

To pull up the available widgets for your phone, hold down your finger on any open space on the Home screen. When you do this, the Home screen "shrinks" and the Widgets icon appears (along with others) at the bottom of the screen (see the two examples in **Figure 5-13**).

OnePlus 9 Pro   *Samsung Galaxy S10+*

FIGURE 5-13

Tapping the Widgets icon opens to a huge array of graphics, games, links, and information bars that you can put on your Home screens.

To add a widget, swipe through the Widget library (shown in **Figure 5-14**) to see all your phone has to offer. When you find a widget you like, with a single option, long press on the widget's icon to place it on the Home screen. Because some widgets have multiple options, the long press doesn't work. But tap the widget icon once, and more options open; you can then select an option by long pressing.

FIGURE 5-14

REMEMBER

If the Home screen grows too crowded, keep in mind that you have *multiple* Home screen pages. Drag the less important short-cuts to those other pages by long-pressing on them and dragging to the right. Voilà! — the shortcuts magically appear on a new page. (Putting them in an organized folder first can help, too.)

IN THIS CHAPTER

» **Making a keyboard decision**

» **Dictating your text**

» **Predictive text and spell-checking**

» **Adding special characters and emoji**

» **Deleting, copying, and pasting text**

» **Printing documents and web pages**

Chapter **6**

Android Typing Tricks with Google's Gboard

You're looking at your new phone, and you're ready to get going. Typing is a major portion of what you do on a smartphone. Sure, you'll also swipe and pinch and pull out with your fingers, but *typing* inputs the data you need. You'll type to search for unfamiliar terms, post to social media, and compose text messages and emails, among other vital tasks. Other chapters help you with specific tasks, but this chapter talks about the physical or — in the case of a mobile device like your smartphone — virtual keyboard.

It's not surprising to realize, as with everything Android, that you have choices. Phone manufacturers often have their own, customized keyboards; Samsung has a clean (simple) one. If the keyboard that came preinstalled on your phone suits you, you have no decision to make. Personally, I've never warmed up to the Apple keyboard, but maybe that's just me. (No, it isn't — trust me.)

In this chapter, you find out about your keyboard options. In addition to the frequent typing that you do on your phone, you may find that you need to print an email message, a message attachment, or a document you find while browsing. I also explain how to accomplish such printing. Let's get started.

Make the Keyboard Decision

After you've opened your phone and you're ready to type something, click to open the browser, and then place the cursor in the text box by tapping on it. You'll see a keyboard pop up at the bottom of the screen.

Your choice of keyboard is quite personal. It depends on many factors, including the size of your finger, the span of your palm, your typing ability, and — most of all — what you're used to. I recommend trying out different keyboards until you find the one that matches your personal typing style.

Selecting a keyboard to use

Before I describe the most popular keyboards, let me assure you that changing the keyboard you want to use is easy. A couple of taps and it changes (or reverts to your favorite). So you can easily try out these various keyboards and decide which one you like best. Here's how it's done:

1. Open or unlock your phone.

2. Swipe downward on the Home screen to reveal the System Settings cog icon at the upper right (or bottom) of the window shade and tap on it. See samples from two different phones in **Figure 6-1**.

3. When the System Settings screen opens, tap on the magnifying glass to make a text box appear (if necessary) and type **keyboard** in the search area.

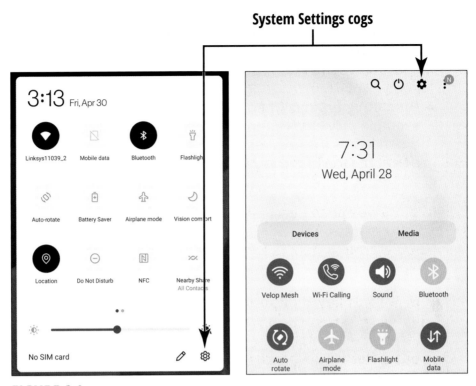

FIGURE 6-1

4. Select an interesting-sounding keyboard from the list by tapping on it.
   Note: If this is a new phone and you haven't added an additional keyboard app, you probably have few options. For example, your choice may be between the manufacturer (such as Samsung) keyboard, SwiftKey, or Google keyboard.

TIP

If you don't see a keyboard you want to try preinstalled on your phone, you can go to Google Play and download it. After you download and install the new keyboard, follow these steps to select it and make it the default.

5. Mark the selected keyboard as the default — for example, by tapping on it — and it will appear the next time you need to type.

Noting keyboard features

You're most likely familiar with the standard *QWERTY keyboard* that appears on a computer or a typewriter, but a smartphone screen doesn't have room for all the keys you find there. The most popular Android keyboards have distinct layouts. Android keyboards also support different options. In my opinion, these are the most significant:

» **Predictive typing:** When you repeatedly type something — for example, "My email address is myemail@gmail.com" — the keyboard learns the text, remembers it for later, and suggests it in a box at the top of the keys. See **Figure 6-2**.

Predictive typing   **Autocorrect**

FIGURE 6-2

Gboard is the most accurate at predictive typing because it learns *slowly over time*, which prevents your phone from remembering misspellings or erroneous data that you entered just once. In my experience, other keyboards tend to learn words too quickly, thereby encouraging typing errors.

» **Autocorrection:** The virtual keyboard, which comes with a pre-installed dictionary, corrects misspellings on the fly.

When I'm texting, I try to remind myself to be careful to reread anything I type before hitting the Post (arrow) key. When I type furiously, I can easily miss an embarrassing (or ridiculous) autocorrection. I find that this problem happens less frequently in

Gboard, which is my main reason for choosing that keyboard over the others.

» **Themes:** Themes change the keyboard colors and can make them beautiful or garish. I must admit that I don't use themes on my keyboard — all I need to see plainly are the letters. A valuable keyboard theme is one that makes it easy to read in strong sunlight. I don't usually type on my phone in strong sunlight, though, do you? (When I do, I just temporarily ramp up the brightness on my phone using the slider.)

TIP

I minimize the brightness on my phone, mainly to prevent eyestrain, by using the slider that appears at the bottom of the Quick Settings pull-down shown in **Figure 6-3**. You can also activate a setting called Eye Comfort Shield (on Samsung) or Vision Comfort (on OnePlus) to filter out blue light to protect your vision further. Hold down the related Quick Settings icon to find options when you want to customize vision comfort.

FIGURE 6-3

- » **Dark mode or Light mode:** Here's an option that's important to many people. If you prefer a dark (black) screen with white text, you can set the keyboard for Dark mode. Black text on a white background is Light mode. **Figure 6-4** shows you how they look. The elusive "they" say that Dark mode is easier on the eyes, but I prefer instead to use the brightness adjustment tools I describe in the preceding tip.

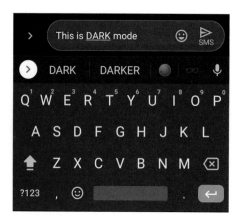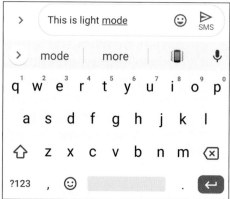

FIGURE 6-4

To set Dark mode, go to the system settings (swipe downward on the Home screen and tap the System Settings cog) and type **display** in the search box. The resulting screen will have the option to turn Dark Mode on or off.

- » **High Contrast Keyboard:** On the Samsung Keyboard, you can access a high-contrast keyboard (shown in **Figure 6-5**), which many people feel is easier to read. Select Samsung keyboard, and from its settings, select the High Contrast Keyboard option.

In a bar above your keyboard, you will see a cog that you can tap to find options to customize the keyboard. Here you can select from a group of color combinations (themes) and other settings. Many keyboards have warnings that say they "may collect sensitive data like credit card data. . ." If that upsets your stomach a little, go into keyboard settings, choose Advanced, and then choose to NOT share usage data.

Settings cog on High Contrast keyboard

FIGURE 6-5

Exploring keyboards you have (or can have)

To see which keyboards are on your phone, go to Settings and search for *keyboard* and then *defaults* or *language*. As you know, Android phones are different from manufacturer-to-manufacturer, but your search for *keyboard* should result in a screen similar to those shown in **Figure 6-6**. If you don't see one you want to try, go to Google Play, search there, and download the one you want.

Notice on the right of Figure 6-6 that, in the Samsung Keyboard settings, you can tap the toggle to have a tiny Keyboard Button icon appear on the screen bottom navigation bar. As you type, you can tap on the tiny keyboard button and a screen like the one in **Figure 6-7** appears.

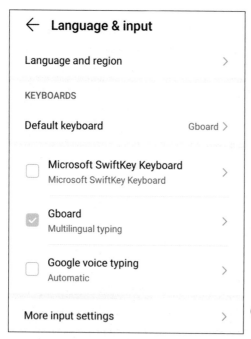

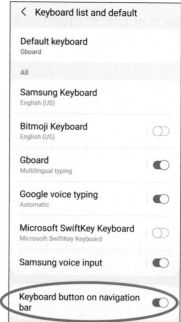

FIGURE 6-6

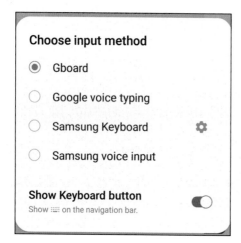

FIGURE 6-7

Check Out Keyboard Contenders

These next few sections give the short story on each of the most popular Android keyboards.

When you download a keyboard (or extra themes) that are not native (pre-installed) on your phone, you will have to agree to another company's "Terms and Conditions," just as you would any new app. Also, be sure to go into System Settings and change the defaults.

Gboard, the official Google keyboard

To start out, I admit that I'm biased. I love Gboard. I install it on every phone I use. It seems other people obviously do, too, because it's had over a billion downloads from Google Play.

Some of my favorite features (which mostly work flawlessly) are described in this list:

» **Voice typing:** I realize that lots of apps support dictation (voice-to-text), but I've found Gboard to be highly accurate.

» **Google Translate within Gboard:** I have friends in other countries whose native language isn't English. Gboard automatically translates your typed text into almost any language you can imagine! The last count was close to 1,000 languages, and you don't have to make any alterations to the phone settings. Just tap the three dots on the keyboard taskbar, and that will give you a translate option. Tap that option, select your language, and type! It's reasonably accurate. **Figure 6-8** shows where I texted a friend in Chinese, as an example.

» **Text correction and predictive typing:** All I can say about this feature is that it just works! It helps speed up your typing and also corrects spelling and increases accuracy. Refer to Figure 6-2.

» **GIF selection:** Unlike with other keyboards, you don't have to download a slew of GIFs to your phone. Google pulls animated GIFs from the cloud based on the search keywords you type, as shown in **Figure 6-9**. GIFs add a bit of spice and fun to your missives. I text a GIF (pronounced like the peanut butter), as shown in the figure, to my adult daughter every day — just because.

FIGURE 6-8

Type keywords here to search GIFs

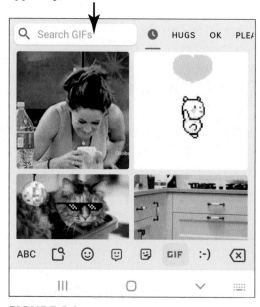

FIGURE 6-9

» **Emoji:** These are tiny icons that can express a multitude of feelings with images. Gboard suggests emoji as you type to emphasize your text — just in case you aren't clear which one to use.

You can also search emoji from the keyboard emoji search. Later in this chapter, in the "Learn the Emoji Language" section, I talk more about emoji.

» **Stickers:** As if emoji and GIFs aren't enough, you can access stickers and cartoons from the keyboard. Some are preinstalled on your phone, and others you will have the opportunity to download.

You can access a slew of other settings by clicking the cog at the top of the keyboard. (Refer to Figure 6-5, earlier in this chapter.) Don't be afraid to try things out. Nothing needs to be permanent — if you don't like an option you've selected, just go back to Settings and deselect the option by tapping on it again. No harm, no foul.

Samsung Keyboard

Samsung has its own keyboard preinstalled on its phones. The official keyboard is proprietary to the operating system of Samsung phones — although you may find many of them using the Samsung name in Google Play. The keyboard, which is pretty amazing, has all the popular options. For example, it has its own preinstalled dictionary and learns unrecognized words as you type them.

The Samsung keyboard has another feature that I adore. When you turn the phone into landscape mode, you see a dual keyboard that makes typing even easier! **Figure 6-10** shows you how it looks in both Light mode and High Contrast mode.

Although I love the keyboard layout, I personally have a problem with Samsung's predictive typing and autocorrect. I tend to type more errors (and have erroneous words replaced) when I use this keyboard.

Samsung also natively supports Bitmoji: You may have noticed these cartoon avatars (see **Figure 6-11**) from friends and on social media. You will need to design your personalized Bitmoji with a separate app, which you can find by downloading the Bitmoji app from Google Play. Using Bitmoji is fun, but I rarely use them, and only on Facebook. Facebook has its own avatar feature.

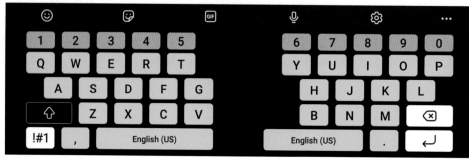

FIGURE 6-10

FIGURE 6-11

Microsoft SwiftKey keyboard

The SwiftKey keyboard, which is available for both Android and iOS (Apple) devices, was so popular on Android phones that the day it launched in the Apple App Store, it had over a million downloads by iPhone users. As of this writing, SwiftKey has been downloaded more than 500 million times. It supports over 400 languages and offers lots of features and options.

Figure 6-12 shows you how the basic SwiftKey keyboard and emoji options look.

 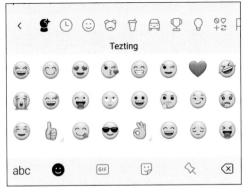

FIGURE 6-12

Speak Words with Voice Typing

I must admit that I feel silly talking to my phone when no one is responding from the other end. But when no one else is around to hear me, I'd rather speak my short, back-and-forth texts. When it comes to longer-form email messages, I prefer typing, because I can produce more formal, thoughtful text.

On Gboard, speech-to-text is precise. To initiate a dictated text message, follow these steps:

1. Tap in the text input area.

2. When the keyboard appears, tap the microphone icon on the keyboard taskbar.

3. Start speaking and you'll see your words appear in the text box.

REMEMBER

Call out punctuation marks as you dictate, as in, "Please pick up some tomatoes **comma** lettuce and green beans **period** OK **question mark**."

4. When you finish, tap on the arrow at the end of the text box, and the text will be sent.

WARNING

If you see a microphone (or other audio) icon at the far right end of the message text box, do not mistake it for the dictation microphone on the keyboard taskbar. That one's for recording and sending short audio messages, as shown in **Figure 6-13**.

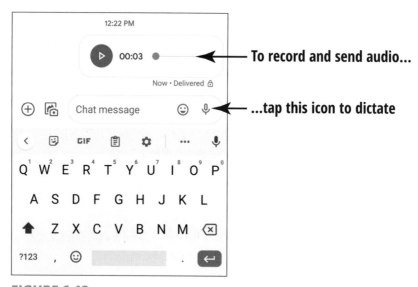

FIGURE 6-13

Spell-Check as You Type

Occasionally (ha-ha!), you may misspell a word. When you do, Gboard has you covered. It suggests spelling possibilities for whatever it thinks you're trying to type. When you see the properly spelled word above the keyboard, just tap the word you want to use, and it replaces your erroneous spelling. Just hope you've selected the correct spelling.

Extended Keyboard and Special Characters

When you look at the basic keyboard, you have access to letters. To capitalize those letters, just tap the up arrow (the Shift key) in the lower left of the keyboard.

To type in continuous CAPS, just double-tap the Shift key, and all the letters you type will be capitalized. (It's just like using the Caps Lock key on your PC). To return to lowercase letters, tap the Shift key again.

To spell foreign words and names accurately, you often need to use a *diacritic* (or alternative character). For example, when you're writing to your friend Françoise, José, or Søren — or using the words *mañana, für,* or *café,* it's nice to use the proper form.

Typing these diacritical characters on a virtual keyboard is much easier than typing them on a PC. Touch and hold down the letter key until you see the variations appear, as shown in **Figure 6-14**. Tap on the one you want and return to normal typing.

Find diacritical characters from your keyboard

FIGURE 6-14

Do you need to type numbers, symbols, or punctuation marks? Just beneath the capitalization key is the **?123** key. (Your key may have a similar, but different, label.) Tap on it, and the keyboard changes to show numbers and additional punctuation marks (on the left in **Figure 6-15**).

If you still need other keys and symbols, you'll be pleased to know that you can find them by tapping on the =\< key. (Again, your key may have a different label.) It appears above the ABC key in the lower left corner.

To return to the alphabet, tap the **ABC** key.

FIGURE 6-15

Learn the Emoji Language

I love using emoji (and maybe a bit too much). *Emoji,* the small, text-size pictographs that you see just about everywhere, are the hiero-glyphs of the 21st century. First created in 1999, they were preceded in 1982 by emoticons, such as :-), typed using the keyboard. You may remember them?

The name *emoji* is derived from the Japanese term for *picture* and *moji* for character. I say they are emoji because, by using one tiny pic-ture, you can communicate all kinds of emotions and illustrate many things. You see emoji everywhere: on clothes and even in interior decorating. I've found them to be fun to use, and I try to keep myself from using them too much.

The Android universe now features 3,521 emojis — and that's way too many for me to remember. You can visit emojipedia.org and take a look at all of them, but it's far easier to use the emoji search that's built right into the keyboard. See **Figure 6-16**.

Type keywords here to search emoji

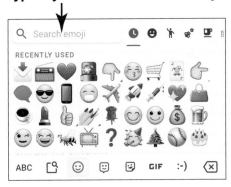

FIGURE 6-16

Tap the Emoji icon and type a word in the search box to describe what you're looking for. As you type, Android suggests emoji for your use. You can also browse emoji, by tapping on the icons next to the search bar.

REMEMBER

Which emoji you have access to depends on your phone's manufacturer and the version of Android that your device operates. Older phones using Android don't have as many options.

If you don't find the exact one you want in a Gboard emoji search, you can type two emoji, one after the other, and see a mash-up, in the form of a sticker to place in text messages.

If you want information about future emoji releases, visit the online authority: blog.emojipedia.org.

Delete, Copy, and Paste Text

If you type as I do, you may want to revise your text after typing and before sending. Revising with the Gboard virtual keyboard is as easy as using the computer's keyboard — only without the mouse.

1. To delete or edit your typed text, simply tap (and hold slightly) in the text box on the word you want to change.

2. The word becomes highlighted, with two tabs; one before the word and one after it. To highlight any other words you want to delete, tap on the tab closest to the word and slide your finger to highlight them.

3. When you've highlighted a word or words in the text box, an action bar with several choices pops up, as shown in **Figure 6-17**.

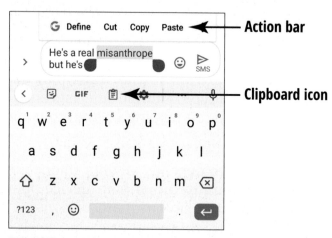

Action bar

Clipboard icon

FIGURE 6-17

This action now gives you the option to

TIP

» **Cut:** Highlight a word or text and remove it. It's placed in the clipboard for future use.

You can see the clipboard's cache by tapping on the small Clipboard icon on the keyboard task bar. Refer to Figure 6-17.

» **Copy:** Any highlighted words are copied to the clipboard.

- » **Paste:** Place the cursor in the text box and tap on Paste. The last text that was placed on the clipboard appears.
- » **Share:** This opens a box allowing you to share to emails, texts, social platforms, and printing.

Then you also see three vertically aligned dots. Tapping on them brings up more options. Here are a few common ones:

- » **Clipboard:** When you tap on the clipboard, you see all previous items you've cut or copied for later use. Just tap on one and it appears in the text space.
- » **Select All:** Choosing this option selects everything in the text box.
- » **Translate:** This is *cool*. You can select from stored languages to translate your words (albeit loosely) into the text area. (I talk about the translation function in the earlier section "Gboard, the official Google keyboard.")
- » **Translator:** Translator, in many cases, may open Microsoft Translator for more robust translations.

And an arrow brings you back to the previous options.

TIP

If you change your mind and decide that you no longer want to send this missive now, just tap on the Back navigation icon at the bottom of the screen.

Print Messages, Documents, and Web Pages

I admit, we're not living in a paper-friendly world, but I still like to print certain documents — maybe a recipe, an email, or an article I've found online that I want to read later. Whatever. I print things. There — call me old-fashioned.

When you subscribe to a newspaper or magazine online, you might want to share an article with a friend or colleague. If you click to share the article via email, the email may have only a link for the recipient to click. Do you know what happens next? When the recipient gets the email and clicks the link to read, they're faced with a paywall asking them to subscribe.

TIP

Super tip! So important! If sharing an online article leads to no access for the share recipient, you can try copying the website link for the URL (which is sometimes different for subscribers) or try this method of sharing: Print the article to a PDF file and attach that file to your email. The recipient can then easily open the PDF on their phone or PC. Problem solved!

Suppose you've opened an email (in this case using Gmail) and want to keep a hard copy of the information in the message. Follow these steps to print:

1. Tap on the three vertical dots in the upper left corner. A list of options appears, one of which is Print. See **Figure 6-18.**

2. Tap on Print, and a miniature version of the page you want to print appears.

3. At the top of this screen, tap the down arrow to select a printer from the existing Wi-Fi printers (assuming that they're on the same network).

   You also have a Print to PDF option, as shown in **Figure 6-19.** If you want to share the email information but don't want to simply forward the message, you can also print the message as a PDF (instead of a hard copy) and share it as an email attachment.

WARNING

Don't forget that you share data with any app you install. Some printer manufacturers require you to install an app, and some don't. As always, after installation, check the app's permissions in the system settings and revoke the ones you don't accept.

FIGURE 6-18

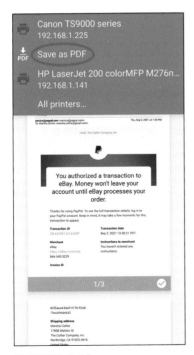

FIGURE 6-19

4. Tap the down arrow for the printer options, including copies, orientation, pages to print, and more.

5. When you've set up every last option, just tap on the Printer icon, and your document should begin to print. You can follow its progress in the Window Shade on the Home screen.

To print from the Chrome browser, tap the three dots in the corner (see? — you're getting the hang of this) and touch Share and then Print. Magical, eh?

Chapter **7**

Handling Notifications and Google Assistant

I f you've ever owned a smartphone, you're probably familiar with the confusion of its messy Home screens. Some Android apps like to place their icons front and center on a Home screen when they're installed, and all end up on a screen in the apps drawer. When I have a new phone, I try hard to keep things orderly. In Chapter 5, I show you how I organize app icons.

Android phones also have the *notifications drawer* (a handy pull-down from the top of the Home screen), which works like a window shade and offers information and access. That is, the Notifications window shade does a lot more than show you notifications from apps. It's a quick launch to a bunch of important utilities.

Because Android phones use tools such as the apps drawer and the Notifications window shade, navigating to find everything on an Android phone can be a challenge to an iPhone user. In this chapter, I show you how to breeze through your phone and manage its content. All those mega- and gigabytes of content you add can fill up space and slow down your phone's operation — if you let your phone run you.

Meet the Android Notifications Window Shade

I realize that *window shade* isn't the official name of the notifications drawer, but that's what I call it. To me, it's much more than a notifications drawer. The way this area looks may vary from one phone manufacturer to another, but it's there, in a similar form. Two window shades (in one) are in the notifications drawer: one that I think is critical to the operation of your phone (the Quick Settings) and the other (notifications) not so much.

Yes, you can receive regular notifications in the window shade, but do you *want* to receive them? The decision depends on how tied to your phone you want to be. Do you need in-depth notice of everything that's happening at the moment? Unless it's a cataclysmic event (or a text message), I'd rather find out on my own time. As newer versions of Android appear, this area seems to remain a stock part of the system. Circles may turn into boxes or buttons, but the controls you tap perform the same functions.

Recognizing notification types and settings

Android notifications come in three flavors: *activities* (like texts and conversations to take action on), *alert* (time sensitive), or *silent* (just to let you know). These terms are fairly self-descriptive because they either appear silently or make your phone vibrate or play a notification sound. You can decide how you want to receive the notification for each specific notifier.

You can access the setting for how any notification behaves by either tapping the down arrow at the right of the notification or just holding your finger on the notification until the options appear, as shown in **Figure 7-1**. In the figure, the notification is set as Other without any sound or vibration.

FIGURE 7-1

TIP

You can also adjust settings for any notifications you receive (or want to receive) by tapping the Settings icon and going to Settings ⇨ Apps and Notifications (or Notifications) ⇨ Notifications ⇨ Advanced.

Disabling notifications (or not)

Android thinks that you want to see all info right now, so, unsurprisingly, notifications may show up when you least expect them. To see the full list, swipe down on the notifications. If you've read one (by tapping on it) and you're duly notified, you can delete the notification by merely swiping it to the right.

You can easily slow the roll of notifications onto your screen — by disabling them. As with all Android features, you have more than one way to accomplish this. You can

» **Disable with a tap.** If you're okay with seeing your favorite team's score on your phone every time they play (I am), that's great. But if you aren't? Tap on the notification (such as the Marlins–Dodgers score notification shown on the left in **Figure 7-2**) to see more information on the topic (as shown on the right).

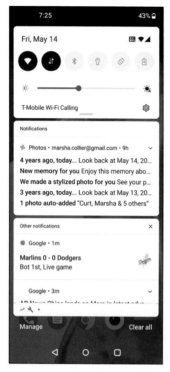
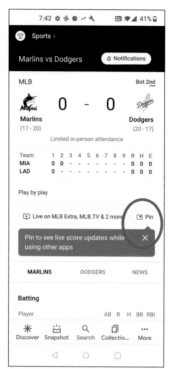

OnePlus 9 Pro 5G

FIGURE 7-2

The screen shown on the right in Figure 7-2 also has a Notifications link in the upper-right corner. When you tap that link, you see a screen where you can either select a different team (see **Figure 7-3**) or just tap the name of the team for which you're receiving notifications. Then it toggles off. Easy! To resume the notification, tap the team's name again to make it active.

TIP

The scores from sporting events are streamed live over Google, so you can check your phone to see the scores in real time. You may also see the Pin button (refer to Figure 7-2). If you tap it, a tiny overlay (shown in **Figure 7-4**) with the game score as it happens appears on any screen of your phone. That is, until you tap the Remove button (in this case) to close it.

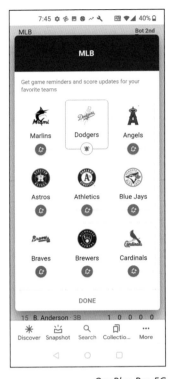

OnePlus Pro 5G

FIGURE 7-3

FIGURE 7-4

» **Disable the miscellaneous.** I had to say *miscellaneous* here because I've received some off-the-wall info that I couldn't care less about. The notifications you see may be from apps installed on your phone or from topics that Google thinks you're interested in (because you searched for them at one time).

Figure 7-5 shows some notifications from my Samsung Galaxy Note 10+. I admit, I do enjoy the Netflix series *Cobra Kai* (I even have the T-shirt), but I really don't feel the need to see news flashes about the show. So I tap the gray down arrow to the right of the notification to reveal two options. When I tap on Not Useful, the notification disappears from the window shade.

Tap here for Notification settings

Samsung Galaxy Note 10+

FIGURE 7-5

REMEMBER

Different types of notifications may have other options. For example, tapping the down arrow on Google News stories may offer the options to save stories or show you fewer stories in a specific category.

Controlling notifications via settings

On the Notifications window shade, you may see the words *Notification settings* at the bottom (refer to Figure 7-5). Tapping this link opens a list of apps, similar to those shown in **Figure 7-6**, that are sending you notifications. From this list, you can toggle apps off to disable their right to send you missives. Of course, click the word *More* or *More Recent* to view more apps that have notified you. Using the Notification Settings link is a more efficient direct shortcut than going through the main Settings cog to Apps (or Notifications).

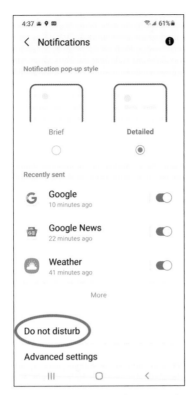
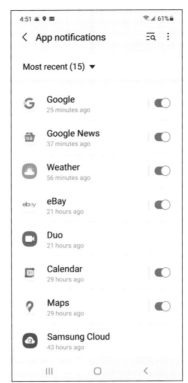

Samsung Galaxy Note 10+

FIGURE 7-6

You can also set notifications to Do Not Disturb. Tapping this option, also shown in Figure 7-6, presents you with time frames to temporarily disable being notified about anything.

Having fun with notifications, or not

An entertaining notification often comes from Amazon (depending on whether it's delivering the package through Amazon Logistics versus another carrier). **Figure 7-7** shows me a notification that my package is ten stops away from being delivered. By tapping the link labeled See Where Your Package Is On a Map, you then see a map of your neighborhood, where you can track the delivery vehicle in real time. This is an amusing pastime for when you really, *really*, want that package.

OnePlus 9 Pro 5G

FIGURE 7-7

I noticed that Instagram took a real liking to my Notifications window shade, but I really didn't need to see these alerts. When I want to play on Instagram, I go to the app itself. For any app you feel this way about, follow these steps to take control over the notifications:

1. Tap the down arrow next to the app notification (in this case, for Instagram). A screen similar to the one shown on the left in **Figure 7-8** appears.

2. Tap the Turn Off Notifications link, and another screen appears (shown on the right in Figure 7-8).

3. Tap the toggle button on the various options to narrow down exactly which app activities you want (or don't want) to be notified about.

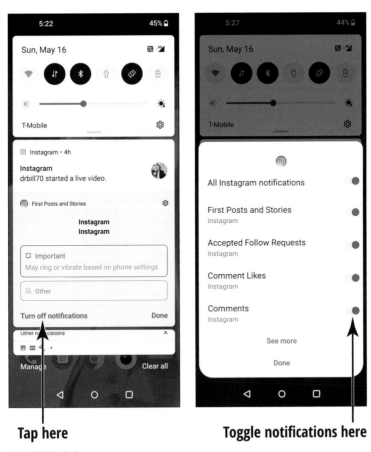

Tap here

Toggle notifications here

FIGURE 7-8

Taming Google Discover news feed

In Chapter 6, I mention that Google Discover is a news feed you access by swiping right from the Home screen on many Android phones. You can also get there by tapping the G logo in a Google search bar. You can access the feed to disable or enable a plethora of notifications, as described in this list:

1. Go to the Google Discover page by swiping right on the Home screen.

2. In the upper right corner of the resulting screen, tap the Circle icon to gain access to your Google account.

If you uploaded a picture of your face to your Google account, that picture appears in the circle. See **Figure 7-9**. If you haven't uploaded a photo and see only your initial, tap the initial in the circle and follow the prompts to set an avatar for your Account.

TIP

After you upload your photo to your Google Account, it appears next to your name in Contacts (and Gmail) within your Android connections.

REMEMBER

In Google Discover news feed, you find lots of ways to customize your Google Android experience. Feel free to tap around and explore. You won't break anything, and you won't change anything unless you toggle it. And toggling goes both ways: on and off.

3. On your account page, tap the Settings link. When the Settings menu opens, tap Notifications.

4. Tap the word *Notifications* again and you see a long list of notifications that you can set to receive or to not receive. (See? I told you it was a plethora.)

5. Review the list and then tap to toggle from off (usually, gray) to on (maybe red or blue). You may find some interesting options here, so why not give them a whirl? You can always toggle off notifications later if they become annoying.

If you've been using Google Search on a phone or a laptop or desktop computer for any length of time, you might be amused by checking out interests on the Discover (Google Account) Settings screen. After you tap Channels & Interests and then Your Interests, Google shows you a list of all the topics it thinks you're interested in, based on your previous searches online.

TIP

Check back with the Your Interests info now and then — it's worthwhile. Google serves up news stories to you based on these "interests." You can review this list and remove topics you don't care about. At the end of the list, you can also add to your Discover list any new topics you follow.

See your face here

FIGURE 7-9

Manage Your Phone with the Window Shade Quick Settings

Other than texting and email, the Quick Settings from the Notifications window shade is the most commonly used feature on my phone. If you want to change almost anything about how your phone functions, you can do it at the top of the window shade, just above the notifications. Again, various phone brands may add special utilities of their own here (as part of the user interface, or UI), but the basics are the same. The next couple of sections illustrate some of the many tools that are available in this area.

The first-up Quick Settings

To access the Quick Settings, unlock your phone and then swipe downward from top to bottom on the Home screen. Your (seemingly endless) notifications appear, and just above them, circles (or rectangles) with icons inside, as shown on the Samsung phones in **Figure 7-10**.

On these Samsung phones, you find today's date and the Settings cog at the top status bar (or bottom, if the phone has a different manufacturer). You may also find other buttons and controls. Referring to Figure 7-10, you find:

>> **Devices:** Tapping this button can take you to either the Samsung SmartThings IoT or Google's IoT. The *Internet of things*, or *IoT*, is an acronym for all the Wi-Fi connected, smart-home devices you may have in your home.

Brightness slider   **Device and Media buttons**

 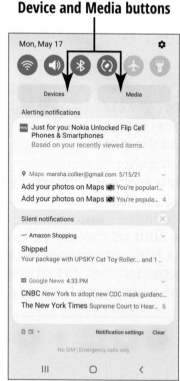

FIGURE 7-10

I personally use generic IoT devices because I can access them from any Android device. The IoT screen points to my Google Nest thermostats (self-learning, smart thermostats) — I have two), a few programable smart plugs I use to turn lights on and off, as well as a Google Nest Hub device that I use as a clock, Google Assistant and Google Photos display.

» **Media:** Tapping this button opens controls to play music on your phone or to share your media with someone via Bluetooth and Wi-Fi.

» **Brightness slider:** Dragging this nifty slider increases or decreases the brightness of the screen. I change my screen's brightness regularly (especially when going outdoors). You can also use the phone's main settings to set up auto-brightness based on ambient light. Truthfully, the auto-brightness (AKA adaptive brightness) never works for me because it often doesn't choose the correct setting for me in any lighting situation.

The full cast of Quick Settings

Swipe downward one more time from the bottom of the first row of the Quick Settings icons (refer to Figure 7-10), and the Quick Settings appear in all their glory — more than can fit on one page, so swipe left to view more settings. A tiny line of dots (shown in **Figure 7-11**) at the bottom of the collection of settings icons are your navigation in this area. The number of dots corresponds to the number of screens of Quick Setting icons you have. The highlighted dot shows which screen in sequence you're on.

In Chapter 13, find out about my favorite quick tools, all of which you can access from the Quick Settings screens. And in Chapter 14, you can find more information about all the magical capabilities that comes supplied with your phone. No external apps needed!

FIGURE 7-11

Get the 411 from Google Assistant

I have to admit that I was slow to adopt Google Assistant. I felt a little silly speaking into my phone and asking it to do something — but no more. In my humble opinion (and that of my iPhone-using husband), asking a question prefaced with "Hey, Google" scores you the quickest and most accurate answer possible because Google Assistant is powered by artificial intelligence (AI). It learns from you and your speech patterns and activates only when you begin your query by saying, "Hey, Google" or "Okay, Google."

To train Google Assistant to respond to your voice alone, swipe down from the top of the Home screen and tap the System cog in the upper right corner. Tap Google (Google Services) ⇨ Settings for Google Apps ⇨ Search, Assistant & Voice ⇨ Google Assistant. Then follow these steps:

1. Under Popular Settings, tap Voice Match and toggle on the setting Hey Google.

2. Tap Voice Model ⇨ Retrain Voice Model.

3. Follow the subsequent instructions to record your voice.

   Google asks you to repeat words and checks off your replies until the assistant is trained.

My Google Assistant doesn't respond to my husband's (or anyone else's) voice. That means a lot, especially when you know you can unlock your phone with your voice.

Knowing what you can do with Google Assistant

You can ask Google Assistant a question like this one: "Hey, Google: What's on my calendar?" Or, while you're listening to music, ask, "Hey, Google: In which year did Cher record 'I Got You, Babe'?" Better yet, right now, ask, "Hey, Google: Play 'I Got You, Babe'" and let's all sing along together!

Sorry, I got carried away (but I did just play "I Got You Babe" on my phone). Google Assistant is my go-to app for almost everything. I send text messages and short emails. I ask to know how many teaspoons in a tablespoon. (Google says three.) When you're traveling, Google Assistant can do currency conversions. Its functionality seems limitless. Google Assistant can even offer up a joke or two, as shown in **Figure 7-12**.

When I asked Google the question shown in **Figure 7-13**, Google Assistant read the answer to me as it opened the screen.

FIGURE 7-12

Installing or deactivating Google Assistant

If you have an older phone and Google Assistant isn't installed, simply go to the Google Play Store and download it to your phone. On some older Android phone models, you need to launch the app by long pressing, then swiping upward from the bottom of the Home screen or by tapping on the Google Assistant icon.

Luckily, if you have a device with Android 11 or higher on it, Google Assistant is right there ready to go. Just be careful about saying "Hey, Google" randomly.

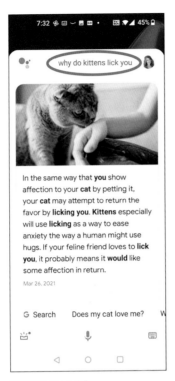

FIGURE 7-13

If you want to turn off Google Assistant, here are the steps:

1. Say to your phone, "Hey, Google. Open Assistant settings."
2. Under All Settings, tap General.
3. Toggle off Google Assistant.

   But why miss all the fun?

3

Let's Start Communicating

Chapter 8

Chatting via Voice or Video

You might believe that writing a chapter on how to make a phone call is unnecessary. I beg to differ! Android has options — so many ways to accomplish even the most basic tasks. Making a call is no different, and the Google Phone app has many settings so that you can customize your experience.

How many times have you jealously watched iPhone users make FaceTime calls with their children or grandchildren — thinking that you had to have an iPhone to do the same? It's a myth — you don't.

Also, maybe you're familiar with the iconic Marimba ringtone on the iPhone? When I hear it, I wonder why anyone would use a ringtone from 2007. (Perhaps they're tech-challenged?)

In this chapter, I show you how to not only dial a call but also change the ringtone, make a video call, and more.

Make a Voice Call

I know that placing (or answering) a phone call seems simple. But doing so may be more than a 1-step process. Here are the two main ways to place a call:

» **Start with your contacts.** You can tap the Phone icon on the Home screen, tap Contacts, and search for the name of the contact you want to call. Of course, using this method means that you already have a set of contacts entered directly into your phone or imported from a previous mobile device. Building your list of contacts is an important starting point.

REMEMBER

If you set a Gmail account as the default storage location for your contacts (refer to Chapter 9 for more info), retrieving contact info is easy as the years go by. When your contact info is safely archived in a folder, you can easily find someone.

If you have a business card to create a contact from, you might be in luck. Today's business card often has a *QR code* printed on it — if you scan this barcode-looking thingy, it may either take you to a website or dial the person's number using your phone.

TIP

If you can't find a QR code scanner app on your handset, open your Phone app. You may see a Google Lens icon at the bottom of the screen. Tap that and a square appears; center the QR code in the square. After the app recognizes the QR code, the predetermined action is activated.

WARNING

Be careful when scanning a QR code to call an unknown (especially international) number. Scams can occur anywhere, and you may end up with expensive charges on your phone bill. As I recommend in Chapter 9, though QR codes can be useful, make sure you know who's on the other end of the connection.

» **Go old school and dial a number.** Tap the Phone icon on your phone's Home screen to see a screen with options to search your contacts, see your recent calls, or dial a number on a numeric keypad. Choose Keypad, tap the number keys to enter the phone number, tap the Phone icon again — and you know what to do from there.

» **Ask your phone**. If you've activated Google Assistant (see Chapter 7 for more information), you can just say, "Hey, Google, call Susan." Google Assistant brings up contacts named Susan and asks which number you want to call. All hands free.

The keypad and call screen design may vary from brand to brand — **Figure 8-1** shows a Samsung layout.

FIGURE 8-1

When you dial a call to a number within the United States from a mobile phone, the area code is required, but there's no need to dial the prefix 1.

Making international calls from your phone

No one ever explained to me the details of calling internationally, so I assume that you likely don't know how this works, either. So, here goes.

All phone numbers in the United States begin with the number 1, and then you dial the area code plus the 7-digit phone number, right? I previously had no idea that the 1 was the international direct dialing (IDD) country code. Every country has its own IDD code number.

On a landline, you dial 011 (the United States exit code) and then the IDD country code (for example, 44 for the United Kingdom) and then the specific phone number. From a mobile phone, you (gratefully) don't have to dial the exit code prefix to make a call out of the country. Now you just need to tap the plus sign (+) and type the country code.

In the case of the UK, I just tap the +, which appears below the 0 on the keypad) and then the country code, 44. In mobile calling, the + (below the 0) replaces the old-fashioned exit code.

REMEMBER

Ask your international friend or family member to text or email the exact number to dial. Request the specific number directly from the source because the number may include a mobile or 2- to 4-digit city code to dial before the actual phone number. The codes vary from city to city internationally and can become *confusing*.

WARNING

International calls aren't included in basic phone plans. Because of the expense, I don't recommend that you call internationally directly from your mobile phone, even if your phone sports a Wi-Fi calling option. In my research, I've seen charges reach as high as $3 per minute, and that's outrageous. Instead, most carriers offer a special international plan (usually, for an additional $15 or so per month). Read on for better money saving long distance tips.

TIP

Though most carriers include international *Short Message Service* (SMS) texting in their regular plans, double-check before you start texting daily with your cousin in Vienna.

Calling internationally with no contract

When you don't have (or don't want to have) international calling service as part of the contract with your carrier, you have other options:

>> The **Google Voice** app is one of my favorites; it's my go-to app for international calls. With Google Voice, you can get a free phone number and a bunch of other features, including video

calling (if the recipient accepts video). I describe other features of Google Voice in the section "Get Voicemail Transcriptions."

- After you set up a Google Voice account, you can make calls from the app with your mobile phone for pennies.

- You don't have to deal with country codes, because the app has a drop-down menu where you can select the country you're calling.

- To find the rates to your calling country of choice, go to `https://voice.google.com/u/0/rates`. I can call family members in Europe for only a penny per minute to a landline or three cents per minute to a mobile phone.

» The **Skype** app has long been a favorite for free app-to-app video calling and chat. It has its own set of unique features, including

- Discounted, pay-as-you-go international (or domestic) calling known as Skype Credit: The service is a bit more expensive than Google Voice. Calls to the UK are 2.3 cents per minute to landlines and 10 cents per minute to landlines.

- You pay a connection fee *unless* you sign up for one of the company's calling plans.

- For more information on the Skype app's discounted calling plans, go to `https://www.skype.com/en/international-calls`.

Check Voicemail

It used to be a simple procedure to check your voicemail: You tapped an icon on your phone and — boom! — you've got voicemail. Not so much now — each manufacturer and carrier has added its own apps and features. This is one reason that you may want to buy a phone (unlocked) from the manufacturer's website: It lacks the carrier's software overlay. (Be warned that manufacturers often add their own apps and features, too. Chapter 2 gives you more information on how this concept works.)

Still, in most Android phones, retrieving voicemail is easy. You may receive a notification on your phone in the Notifications window shade. Unlock your phone, swipe downward on the Home screen, and tap voicemail notification from the window shade.

If you have no visual notification and you need to call your voicemail, follow these steps:

1. Tap the Phone icon on the Home screen. **Figure 8-2** shows that different brands have different icons. The traditional Android Phone icon is blue.

FIGURE 8-2

2. On the resulting screen, you may see the dialing keypad. (Refer to Figure 8-1.) If the screen you see is a listing of calls and you find no hot link to Voicemail, look for a keypad link and tap that.

3. Hold your finger on the number 1, under which is an icon resembling a tiny reel-to-reel tape recorder (the kind, shown in **Figure 8-3**, that used to dwell inside answering machines). Your phone should call your provider's voicemail system.

   That's the direct line to voicemail in most phones. You shouldn't need an extra app.

FIGURE 8-3

Get Voicemail Transcriptions

You *can* set up your phone to transcribe your voicemail and have it appear as a notification, an email, or in the Google Voice app without having to download any extra apps. When you buy a phone directly from a carrier, someone might have disabled the option to use the native Android feature so that you then use *the carrier's* app.

Although the carrier's app may let you read transcribed voicemail for free for the first few months, that may be just a trial period to get you hooked on the ease of reading voicemail.

1. Tap the Phone icon on the Home screen to open the Phone app.

2. In the upper right corner of the Keypad screen, tap the Menu icon (usually, three vertical dots) and then tap Settings on the drop-down menu. (See **Figure 8-4**.)

FIGURE 8-4

3. Tap Voicemail on the resulting screen and then Visual Voicemail.

As shown in **Figure 8-5**, my T-Mobile Visual Voicemail wasn't set up, so I tapped the Call Voicemail link. I had no voice messages but listened to the default you-have-no-messages message and then disconnected.

I've used Visual Voicemail previously on my Google Voice account (with my Google Voice number) and in T-Mobile's Visual Voicemail app. I asked my trusty editor, Leah, to leave me a voicemail message so that we could see what happens. Son of a gun! The message showed up in Notifications window shade, in my Google Voice app, as shown in **Figure 8-6,** and in an email from Google Voice.

REMEMBER

Google uses supercomputers to transcribe these voicemails, but they aren't connected to you or your account.

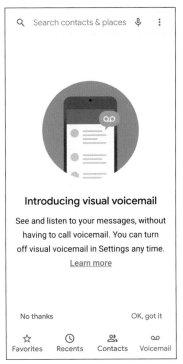

FIGURE 8-5

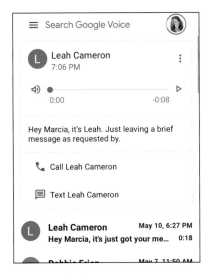

FIGURE 8-6

Reply to an Incoming Call by Sending a Text

I never fail to receive an incoming phone call at the wrong time: My phone invariably rings when I have to either sign for package delivery or grab an available seat in my doctor's busy waiting room. Phones ring at the most inopportune times, and answering the call often isn't a comfortable option.

You may sometimes let voicemail do the work of taking the call. At other times, you want the caller to know that you know they called and that you'll get back to them at a more convenient time. To do so, you can reject the call and send a prewritten text message without making noise or distracting anyone.

Automatic text replies go by different names. Android offers Quick Response texts (or Quick Decline messages, on Samsung) SMS texts that you can send with a tap of your finger.

When a call comes in from a known contact, you see a notification in one of two locations, depending on whether your phone is locked, or unlocked but otherwise occupied.

TECHNICAL STUFF

When I'm at home, I always connect my smartphone to my home Internet, so my voice calls show up as *Wi-Fi Calls*, as you may note in some of this book's screen shots. Wi-Fi calling relies on more than the cellular network alone. If the connection benefits from using Wi-Fi, your phone automatically switches to your home Wi-Fi. You can find the setting to enable Wi-Fi calls in the main Settings cog, under Network (or Connections) ⇨ WIFI Calling. (It's often turned on by default.)

If your phone is unlocked, you may see (at the top of your phone) a message similar to the one shown in **Figure 8-7**. To respond to the call with a text, follow these steps:

1. Tap the notification area showing the caller's name.

2. On the resulting incoming call screen, you can reject the call but still reply and send a Quick Response message, as described in the next step list.

FIGURE 8-7

When your phone is locked, the process is easy. When you see the notification that a call is coming in, follow these steps:

1. Tap the Reply link that appears in the middle of the screen, as shown in **Figure 8-8**. A pop-up menu showing your preloaded responses appears.

2. Tap the response that fits the situation. The call is rejected, and the SMS (text) message is sent to the caller.

FIGURE 8-8

Android phones come with prewritten texts for these situations. You can find the responses by opening the settings in your Phone app. Tap the Phone app icon and then the three vertical dots in the upper right of the keypad screen. Then tap Settings ⇨ Quick Responses (or Quick Decline Messages).

The common generic responses are shown on the right in Figure 8-8. Some phones say *Quick Decline* or *Reply*; when you choose Reply, you can add your own short custom response.

You can edit these responses either while you're responding to a call or at your leisure, by visiting them in the settings. I suggest that you change or add responses when time is not of the essence.

Set a New Ringtone

Your phone's ringtone is a personal choice. It says a lot about you when other people hear it, and it can evoke an emotional response for you when your phone rings. I'm never happy with the default ringtones. I find that many are downright annoying. The Marimba default ringtone on the iPhone comes to mind — it just makes me shudder.

Using a built-in ringtone

Good for you that your phone comes with a decent variety of ringtones from which you can choose! Personalizing the ringtone may be something you want to do. Just follow these steps:

1. Swipe down on the Home screen and tap on the main Settings cog in the upper right of the pull-down window shade.

2. On the resulting screen, tap on Sounds and Vibration. Then tap on Ringtone. You should see a long listing of stock ringtones for your phone, as shown on the right in **Figure 8-9.**

3. Turn up your phone's volume and tap each ringtone (one at a time) to select and play it.

4. When you find a ringtone you like, leave it selected and tap the arrow in the upper left to go back to the Sounds and Vibration screen.

Downloading a custom ringtone

You can also go for a custom sound. If you want a custom ringtone, like a few of my friends use (check out the nearby sidebar "Friends and the ringtone craze"), the setup is a little more involved — but not much. Most ringtones are MP3 clips lasting not even 30 seconds, or you can use a song you've downloaded to your phone from Google's YouTube Music or from Amazon Music.

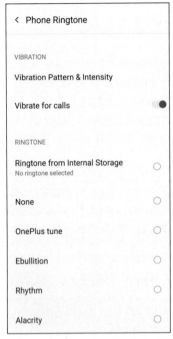

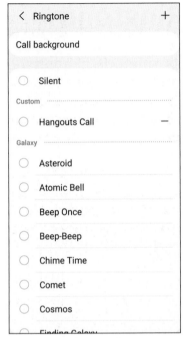

OnePlus 9 Pro 5G

Samsung Galaxy S10+

FIGURE 8-9

As an alternative, you can search the Google Play Store for *ring-tone* and download an app you find there. Or go to the Zedge website `www.zedge.net` using your smartphone's browser and search for ringtones from there.

WARNING

I need to say this more than once: When you pick an app from the Google Play Store and it asks you for permissions that make you uncomfortable, *don't download it.* Why would an app need the rights to modify your device settings? Or read or modify your file storage data? Or use your camera and microphone? Although you can remove some of these permissions in the app Permissions in the main Settings, some apps don't work unless you approve blatantly unacceptable terms. Beware. *Caveat emptor.*

FRIENDS AND THE RINGTONE CRAZE

Back in the early aughts, when I first started using my mobile phone, I was all about ringtones. (Yes, I also had a little charm hanging from my phone.) Ringtones at that time cost as much as $2.50 each — but downloading a full MP3 song cost only 99 cents. According to AndroidAuthority.com, by 2004, ringtones were a $4 billion industry.

In those days, I downloaded the Zedge app on an early-model Samsung Galaxy. The app opened up a world of new ringtones. Believe it or not, Zedge is still around and is a publicly-traded company. The app is well on its way to scoring nearly half a billion downloads. Even though custom ringtones are on a down-trend, lots of people are still using them.

For example, my husband set his ringtone to the theme from the 1966 movie *Our Man Flint.* One of his friends set their ringtone to the intro to the *Batman* theme. Another has the *Star Wars* stormtrooper's intro, and yet another has the *Benny Hill* theme. My daughter goes old school with Scott Joplin's *The Entertainer* — do you remember that one, from the movie *The Sting,* starring Paul Newman? One friend uses the *Addams Family* theme for an incoming call from his brother-in-law because the guy looks like Uncle Fester.

Zedge has an app in the Play Store, but if you want to download a ringtone from the Zedge website, follow these steps:

1. Type `www.zedge.net` on the address bar of your phone's browser, tap Go, and then choose the Use in Your Web Browser link.

2. On the Zedge website, tap the Ringtones link and then search for a musical ringtone by name. I searched the website for "The Entertainer" and found many versions of it.

3. If your search results show multiple ringtone choices, tap them to hear different versions. After you find the version you want on the website, the rest is easy.

4. Tap on the version you want to download, and a new screen opens. You're asked whether you want to continue in your browser or down-load the app.

5. Click the browser option, and a screen offers another prompt to download the app, along with a countdown clock that tells you to wait ten seconds (shown on the right in **Figure 8-10**).

Just wait it out and your ringtone downloads!

FIGURE 8-10

Activating a downloaded ringtone

Now, when you go to select a ringtone on a pure Android (Pixel) or OnePlus phone, you see an option to add a ringtone from internal storage (refer to Figure 8-9). Ta-*dah!* But in my testing, changing the ringtone doesn't work in this straightforward fashion on a Samsung phone.

For a Samsung phone, follow all the steps to download the file (see the earlier section "Downloading a custom ringtone"). Then follow these steps:

1. Swipe down on the Home screen and tap the Settings cog in the upper right corner.

2. Tap Sounds and Vibration ⇨ Ringtone. On the resulting screen, you should see a plus sign (+) in the upper right corner of the list of ringtones.

3. Tap the plus sign, and then give Samsung's Sound Picker app permission to gain full access to all files in your phone's storage (see **Figure 8-11**). On the Sound Picker screen, you can select and play the custom ringtones you downloaded.

REMEMBER

Click the option to remove permissions if the app isn't used often, as shown in Figure 8-11. Forever is a long time — if the Sound Picker app isn't used for a few months, the permission is removed.

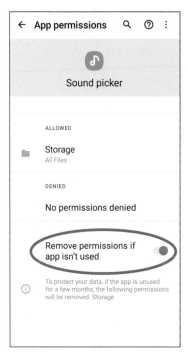

Samsung Galaxy Note 10+

FIGURE 8-11

4. Select the custom ringtone you want to use and then tap Done. You return to the list of ringtones with your custom selection now in the list.

5. Tap the back arrow in the upper left corner of the Ringtone screen. Your custom ringtone should now be active.

Spend Face Time with Family and Friends

No, I'm not suggesting that you use the FaceTime app — that's Apple's trademarked property. I'm talking about *face time*, like looking at, and speaking to, the person on the other end of the call. Google has two apps to help you score some face time.

Google Duo

The most user-friendly video call app is Google Duo. This app is *simple* to use — it's just like making a voice call.

1. Download the Google Duo app from the Play Store, if it's not already installed on your phone.

2. Tap to open the Duo app and grant it permission to gain access to your camera. I figure that's a requirement for video calls, so tap Give Access.

3. On the resulting screen (as shown on the right in **Figure 8-12**), tap While Using the App. Choosing this option limits permissions.

4. Tap the option to allow Duo to record audio. Record? Hmmm. Limit the recording permission to While Using the App.

   Now, for an amusement break, Duo has several filters and funny effects to try out. (**Figure 8-13** shows one of my favorite effects in a screen shot of me talking to my editor.)

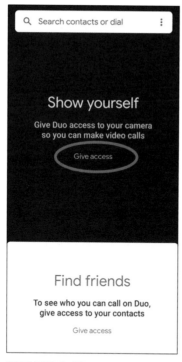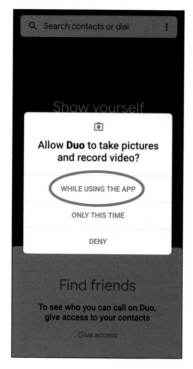

FIGURE 8-12

5. After you finish playing with the effects, tap the *X* in the upper left corner of the screen and return to the app.

6. Tap to give access when Duo asks permission for access to your contacts. Because — let's face it — you can't call anyone if Duo can't access your contacts.

Now you get a chance to make your first Duo call — or not. You can close the app and select a contact to call later.

REMEMBER

To make a call in Duo, both you and the contact you're calling must have Duo installed on your phones.

FIGURE 8-13

Google Meet

No doubt you've heard about the famous videoconferencing app Zoom. *Google Meet* is Google's super-robust response to the infamous app that has forever changed business communications.

If you have a Google account (you do), you can hold Google Meet video meetings with as many as 100 people, for as long as 60 minutes — for free! One-to-one calls with Google Meet are always free.

When you need to contact a group, you can set up a videoconference from the app on your phone or from a PC. The other person or persons *do not* have to have Google Meet installed anywhere, because participants enter Meet by way of a shared URL.

I find Meet especially useful on a PC (either a laptop or desktop) because of all the bells and whistles it offers. You can find more information about Meet at `apps.google.com/intl/en/meet`.

Chapter **9**

Keeping Track of Friends and Appointments

These days, you never need lose a contact's information. With use of the cloud, all data exists *ad infinitum* (purportedly). The definition of *the cloud* is complex, but for keeping information (such as your phone and email contacts), it's a remote place where data goes to be stored. The cloud is basically a metaphor for the Internet — or a "cloud" of computers, storage servers, and software. (Whatever.)

Rather than a physical connection to the storage computer, web-based and mobile tools (or *apps*) connect via the Internet to a central location. As long as you can access the Internet, you can retrieve your data. In addition to finding contact information, you may want to set up your phone as the go-to calendar for your important appointments and events. That information is stored in the cloud (by way of the Calendar phone app) as well.

In this chapter, I present instructions on how to compile data for your phone's contacts and keep that data safe and available. Also, I help you take advantage of the helpful Android apps — for contacts and calendars — that are available to you from your mobile phone. Long story short: If you use the right apps, you can access all the data you've chosen to store in the cloud.

Establish Your Phone's Contacts

I have plenty of faith that the major phone manufacturers will be in business for a long while and will maintain a record of your contacts' information. Problem is, what happens when you want to change phone brands? Can you easily transfer your contacts? Processes can change quickly in the tech world, and you never know what can happen. Check out the nearby sidebar "Where are my contacts?" for a look at my experience with evolving platforms for contacts.

WHERE ARE MY CONTACTS?

Some years ago, I knew by heart the phone numbers of my friends, business associates, and family members. But I also had a hard-copy address book with alphabetized tabs so that I could flip to less familiar contacts quickly. Every entry was hand-written in a different ink color. Sound familiar?

I completely lost the need to memorize addresses and phone numbers when the Rolodex hit my desk. It seemed like a perfect storage solution.

But when digital tools became the number-one choice for tracking contact information, everything changed. I was using Outlook on my PC, but as I moved to mobile platforms, I moved to the Gmail and Google Contacts apps. But (big mistake) I forgot to import my old contacts list; luckily, I had printed a paper copy — which I still refer to.

Person by person, I move contact information into my digital platform. But I encourage you to use the information in this chapter to see how to avoid making my mistake.

Starting out right with Google Contacts

Two things you can bank on: iPhones can always store data with Apple, and Android phones can always store data with Google. I recommend that you use the Google Contacts app to safeguard the contact information on your phone. In Chapter 3, you find out how to set up the Contacts app and have it available for use when you're texting, emailing, or calling.

If you're not planning to use Google Contacts as your main contacts database, you can skip the next section. But when it comes to technology, you can't have too many backups.

Importing old address books

So that you don't fall into the Outlook mess I mention in the earlier sidebar, "Where are my contacts?" it makes sense to import contacts into a main list from your other email accounts. Because this book talks about Android smartphones, it also makes sense to establish Gmail as the main list. Many people use different email addresses for various groups, like family members, religious organizations, or schools. I recommend importing contacts from these other addresses so that you have all your contacts under the Gmail umbrella.

Gmail falls under your Google Account which ultimately gives you control over all the Google products — from Gmail (email), Voice, and Contacts, to Documents, Spreadsheets, YouTube, and anything else Google owns (Duo, Meet).

Keep your other email accounts open. You know the ones — Yahoo, Hotmail, and Outlook. They come in handy as alternative email addresses for security confirmation purposes — or for using on websites that you plan to never visit again. If you close out and abandon your email address, it may be recycled and assigned to someone else. In this situation, you might experience problems if a confidential email from a long-lost friend arrives in a stranger's mailbox. Also, a recycled address can open the door to an impersonator or identity fraud.

The most secure way to import contact information is to access your Gmail account from a web browser — I used a PC to follow the steps, but you can also use a tablet that can access the web version of Gmail through the browser:

1. Go to mail.google.com and sign in to your Gmail account. If you are using (and I hope you are using) 2-factor verification, Google sends a code to your phone to confirm that you are — absolutely, positively — you.

2. Click the Settings cog in the upper right corner. The Quick Settings sidebar appears.

3. Click the See All Settings button at the top of the Quick Settings options list. A daunting number of settings appear, but — luckily — they're segmented into categories.

4. Click the Accounts and Import link, as shown in **Figure 9-1**.

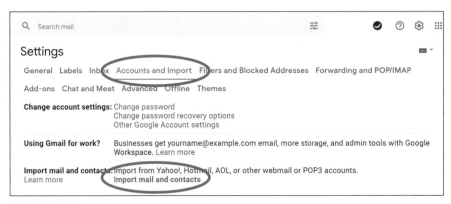

FIGURE 9-1

5. Click the Import Mail and Contacts link (which is, ironically, in the Import Mail and Contacts section).

6. In the pop-up window that appears, type the email address at your other service and click Continue. See **Figure 9-2**.

7. In the next window, you're notified that your data will be unencrypted during the transfer. Click Continue.

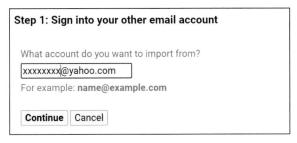

Step 1: Sign into your other email account

What account do you want to import from?

xxxxxxxx@yahoo.com

For example: **name@example.com**

Continue   Cancel

FIGURE 9-2

8. In the resulting Sign In window (in this example, from Yahoo), sign in to your email account. If you have 2-factor authentication set up, your email service sends a code to your phone for you to confirm online.

9. After reading the privacy information for each item (by clicking it to reveal it), click the Agree button, as shown in **Figure 9-3**. You see a notification that authentication has been successful. Click the *X* in the upper right corner to close the window.

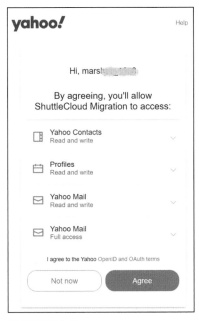

FIGURE 9-3

10. Select check box options to mark what you do (or don't) want to import. These options appear in a small window, as shown in **Figure 9-4**, where you can import just your contacts — that's the plan — or all of your email. Click to remove any of the prepopulated check marks you don't want, and then click the Start Import button.

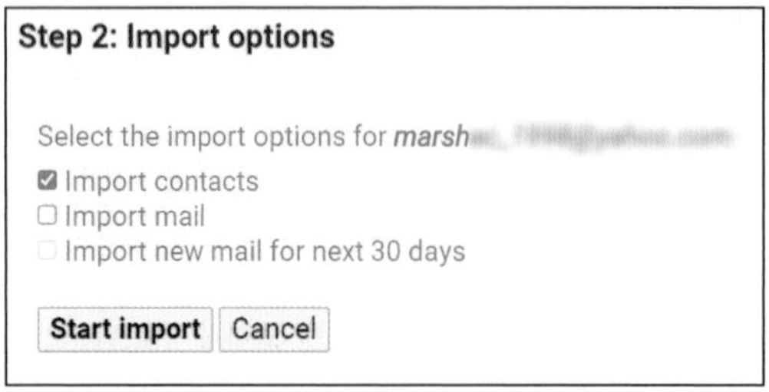

Step 2: Import options

Select the import options for *marsh_____yahoo.com*
☑ Import contacts
☐ Import mail
☐ Import new mail for next 30 days

[Start import] [Cancel]

FIGURE 9-4

You can follow a process similar to the one in these steps with any of your other email accounts where you've stored contacts or want to import old email. After you import all the contacts' email addresses from your other sources, they become part of your Google Contacts list.

Save Contacts from Email

Saving contact information from email messages you receive is a simple and easy way to build your contacts list. When you meet someone new and they send you an email, their name appears at the top of the message.

Open the email and tap the small down arrow next to the words *to me* (see **Figure 9-5**). A drop-down text box appears with the sender's email address, your email address, the date and time the email was sent, and whether the email was encrypted.

FIGURE 9-5

From the drop-down text box, you can make this sender a new contact or add this new email address to an existing contact by following these steps:

1. Long-press (press and hold versus tap) the email address to highlight it. A menu with commands pops up and gives you the options to send an email, copy the email address, or share the email. See **Figure 9-6**.

   Make the long-press on the message a *soft* long-press — or else you'll open a brand-new email to that person. If this happens, tap the three dots in the upper right corner to discard the email.

Tap the three dots   **Then tap to add a contact**

FIGURE 9-6

2. But wait — there's more! Tap the three magic dots at the end of the pop-up menu (after *Share*) to find the option to add a contact.

3. Tap the Add button (on the right in Figure 9-6) to open a new window (shown in **Figure 9-7**) with a couple of options. You can

   - **Add this email address to an existing contact:** Type the contact's name into the Search Contacts field and then tap the magnifying glass to find the contact's data.

   - **Tap the plus (+) sign:** This option creates a new contact.

4. On the resulting Add to Contacts form (shown in **Figure 9-8**), type the contact's first and last names where prompted. If you have more information about the contact that you want to add, tap the More Details link for lots of extra options.

Add to existing contact

🔍 Search contacts

\+   Create a new contact

FIGURE 9-7

FIGURE 9-8

REMEMBER

On a Samsung smartphone, this process of adding a contact may look a bit different because Samsung has its own Contacts app. But the process for moving contacts from Gmail and adding them to Gmail is the same. In most versions of Samsung Contacts, you can select to make Google your "Default" contact storage area, so when you put a contact in your Samsung device, it will sync it to Google.

Add a Contact in Other Ways

Sometimes you give your number to someone, and they text or call you one day. You may or may not have picked up the call, the person may or may not have left a voicemail, but when you look for the number on a search engine, you see that it belongs to someone you know. Whether you receive a text message or a phone call, you can add the sender to your contacts.

Importing contacts from texts

When a text message comes in and the number isn't tied to a contact on your list, all you see is the phone number. Hopefully, after a bit of back-and-forth, you find out that this is a long-lost friend (or your doctor's office).

To add a contact from a text message, follow these steps:

1. Tap the text to open it. You may see a message identifying the person. If so, it's easy from here.

2. If you know who sent the text, tap the Add Contact link (see **Figure 9-9**) and follow the steps for adding a contact from email. You can see them in the earlier section "Save Contacts from Email."

FIGURE 9-9

Adding a contact from the call log

An incoming call without caller ID is annoying, but you may find that the phone number belongs to one of your friends. If you do, just *soft-press* the number (a tap that's too heavy initiates a phone call) and a submenu appears. As shown in **Figure 9-10**, all you need to do is tap the Add Contact option.

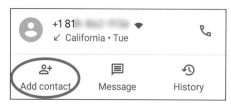

FIGURE 9-10

Type In a Full Contact in the Contacts App

Somehow, adding a contact directly in your contacts app is *so* satisfying. The Google Contacts app gives you room to record *all kinds* of information including birthdays and notes. You can tap the Contacts app icon on either the Home screen or a screen in the Apps drawer. **Figure 9-11** shows the official Google Contacts app icon (on the left) along with the icon for the Samsung Contacts app (on the right).

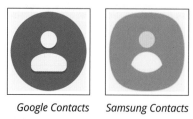

Google Contacts *Samsung Contacts*

FIGURE 9-11

Some phones, like the OnePlus, have their own icons for Google apps. You can always verify whether an app is from Google: Open the app, tap to find the Settings, and look for the Help & Feedback link. When you tap it, an official Google app takes you to a Google help page.

Note: If you intend to make the official Google Contacts app the default, I recommend you set this app as the default from the main Settings cog, which you access from the Quick Settings.

When I tap the three horizontal lines (the Hamburger menu) in the upper left corner of my Google Contacts app, I see the menu shown in **Figure 9-12**. This screen shows that, in the past, I have *imported* (restored) contacts from other phones to this new one.

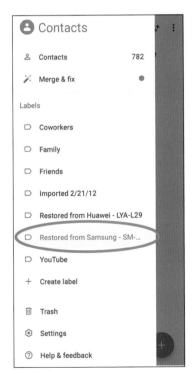

FIGURE 9-12

To enter a new contact directly in the Google Contacts app, follow these steps:

1. Tap the Google Contacts icon to open the app.

2. In the lower right corner of the resulting screen, tap the plus (+) sign in the blue circle to access the Create Contact screen, as shown on the left in **Figure 9-13**.

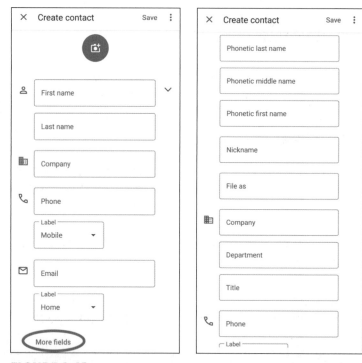

FIGURE 9-13

3. Type any information you have and then tap More Fields. The right side of Figure 9-13 shows the first screen you see as a result, but the app has about four pages of fields where you can save information, such as

   - Additional email addresses and phone numbers
   - Physical addresses
   - Birthdays

The form is *extensive*. At the end, you can tap Add Custom Field for extra data.

4. Don't forget to tap Save when you're finished.

TIP

If your phone gives you any trouble when entering a contact, you can also go to contacts.google.com and, as a backup method, add your contact's information via your PC.

Merge Duplicate Contacts

You may enter duplicate contacts because you receive unique information about a person at different times. Or duplicates can show up whenever you import a contacts list from another email address or device. You can fix this problem!

Refer to Figure 9-12 and notice a small red dot next to the Merge & Fix option. The icon to the left of it looks like a magic wand because, like magic, Google Contacts found these errors.

Tap the Magic Wand icon and you see a menu, shown in **Figure 9-14**, with (in this case) two options:

» **Keep Contacts Up to Date:** This option suggests that Google Contacts has found more information for eight people I've contacted.

This is where the fun (and the magic) happens. Tap the Keep Contacts Up to Date option, and then you have the choice to dismiss or accept each one. (I dismissed only one, an email address that I can't unsee and don't want to be reminded of.)

» **Add People You Email Often:** Google Contacts suggests that I add people whom I email often but are not in my contacts. (I see 19 of them — yikes!) Tap this option to deal with the folks who aren't on your contacts list.

In case you're interested, the 19 people Google found whom I email (but are not on my contacts list) are people I'm feeling a bit iffy about. I just backed out of that screen to deal with it later.

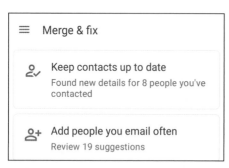

≡ Merge & fix

Keep contacts up to date
Found new details for 8 people you've contacted

Add people you email often
Review 19 suggestions

FIGURE 9-14

Customize, Delete, and Update Contacts

Selecting an individual contact in the Contacts app offers options to adjust the person's priority or information. Open the Google Contacts app and tap the contact whose info you want to alter. Then proceed with changes:

1. Tap the Edit Contact link in the lower right corner. On the resulting screen, you can update the person's contact information. Tap Save in the upper right corner to save your changes and return to the contact's info screen.

2. Tap a hollow star (☆) near the upper right corner to make it a solid star (★). A solid star indicates family members or close friends who will receive preference in contact listings or quick dial.

3. Tap the three vertical dots in the upper right corner to open a menu, as shown in **Figure 9-15**.

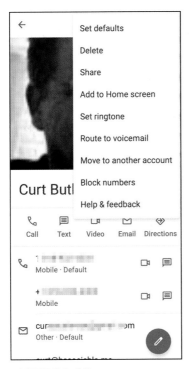

FIGURE 9-15

Here are the tasks you can perform from this menu:

» **Delete:** Tap here to follow the steps and delete the contact from your list forever.

» **Share:** Tapping Share allows you to select which information on this contact you want to share with someone else via a .vcf file. You can add this tiny file to a text or email message, and when the recipient clicks the file, the shared contact's information is imported into the recipient's contacts list.

REMEMBER

This sharing function in Google Contacts has a helpful privacy feature built in. To help protect your contact, you can control how much information that you choose to send in the .vcf file. A screen pops up where you can add (or remove) checkmarks next to the data fields you want to share. The Samsung Contacts app shares the entire contents of the file.

- » **Add to Home Screen:** This option sends a shortcut to your Home screen so that you can just tap the icon to connect with your loved one.

- » **Set Ringtone:** I love this one! In Chapter 8, I talk about ringtones and where to find them. Tap here to choose the ringtone that plays whenever your partner (or another contact) calls.

- » **Route to Voicemail:** Tap this option to have every call this person makes to your number go directly to your voicemail.

- » **Move to Another Account:** If you have control of other Google accounts, you can move a contact from one to another.

- » **Block Numbers:** When this person calls or texts, they cannot ring your phone. (You can also block a contact from your phone app or call history). The person's calls can proceed to voicemail, but they receive no notification that you have blocked them. Their text messages to your phone won't go through. A Delivered notification is never delivered on their end.

- » **Help & Feedback:** Tap here to access Google's Help pages if you get lost.

Energize Your Calendar App

Google Calendar app is a powerful tool. I add just about every event, appointment, expected delivery, and deadline to my calendar. Because I have a Google account for all my devices, the calendar appears on my phone, my tablet, and my Google *Nest Hub* (a smart display where I can access devices and photos linked to my Google account) at my bedside. I'm always checking the calendar (even to see when I have to cancel that HBO trial I signed up for).

Figure 9-16 shows my calendar for a month in the future (or at least it was on the day I wrote this chapter). Notice that some calendar entries carry over from week to week or from year to year. To add an entry, just tap the multicolored plus (+) sign in the lower right corner and follow the steps.

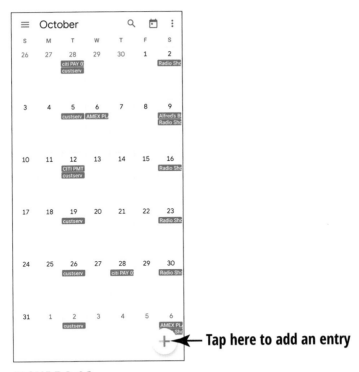

Tap here to add an entry

FIGURE 9-16

TIP

You can visit the browser version of Google Calendar at `calendar.google.com` on a PC or tablet for a far more robust experience. At the end of the year, you can also print a calendar to store for tax purposes. I do this every year.

WARNING

Be careful! Your phone's manufacturer may have installed its own calendar app in place of the standard Google version. I have, in the past, accidentally opened the wrong calendar on a phone and added an appointment. Unfortunately, that appointment did not show up on Google Calendar, and I missed the call. Mobile manufacturers' calendars can sync with Google and import the Google events — but it doesn't always work in reverse.

I've found that the more items I add to the calendar, the more organized I am. In Chapter 13, I show how handy it is to add an address to every appointment. When you do, you can find directions, view a traffic report, and receive a notification, which helps you before you even head out for the appointment.

Add Calendar Events from Gmail

You can also add events and appointments by way of incoming Gmail. Anyone can add an appointment to their own calendar — and add your email address as an attendee. After the event is saved to their calendar, you receive a Gmail message letting you know about the event.

Figure 9-17 shows an email invitation from my husband for a dinner date. After I tapped Yes (indicating that I would go), Google automatically added the dinner date to my calendar.

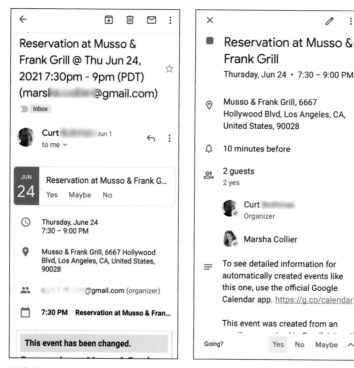

FIGURE 9-17

TIP

This automatic entry also works well with airlines and hotels. When you receive a confirmation email (in Gmail) and accept the invitation, the event automatically is added to your calendar.

Chapter **10**

Texting with Poise and Character

You're probably lucky that you're not on my texting list. I look at texting as not only a method of written communication but also a form of sharing emotions. I prefer texting to almost any other form of communication because the message arrives instantly and — if it isn't urgent — can be answered later.

Texting is a main function on your Android smartphone, and many people use texting daily. You accomplish *texting* — typing messages to send to specific recipients — by using one of any number of messaging apps. A primary message app comes preinstalled on your phone, but you have other options. And you can use texting for a whole lot more than asking "What's for dinner?" or confirming that you'll "meet at 57th and 3rd." You can also share GIFs (small, animated files, pronounced "jifs"), emoji (letter-sized, expressive icons), audio, photos, videos, links from the web, and even files (usually limited in size).

In this chapter, I tell you how to take your texting to a new level with options for composing, enhancing, and sending text messages. It's time to up your texting game! Impress your friends and relations with your newfound skills.

Compare Types of Text Messaging

As you may know, the Apple iPhone has its own walled garden for enhanced texting in iMessage over cellular networks and Wi-Fi. The messaging is specifically designed to be used from iPhone to iPhone — and less so to Android phones. Google's Messages app sends messages to all phones, in the same way.

The type of text you send is based on what you're sending via Messages and on which operating system (OS) you're using. You may have heard initialisms like *SMS* and *MMS* regarding text messaging and wondered what they're all about. These new terms represent text *messaging protocols*, which are sets of technical rules governing how the messaging works. Check out the later sidebar "A brief history of texting" for info on the origins of texting as we know it today.

The following list shows milestones in the evolution of texting protocols that are in use today:

» **Short Message Service (SMS)** was the first and most basic form of texting, limited (as you may remember) to 160 characters per message.

» **Multimedia Messaging Service (MMS)** arrived in the early 2000s, allowing you to attach small photos and very short videos to your messages.

» **Rich Communication Services (RCS)** is the newest iteration of texting for Android. It added the ability to send large files and add animations and interactive media.

To go even further, Chat — an add-on to RCS — is part of the pre-installed Google Messages app. With Chat, you can add others to a group conversation, see (by way of a small animation) when others are typing, share higher-quality photos, let recipients know when you've read their messages, and benefit from end-to-end encryption. To take advantage of these features, all (or both) participants in a conversation must have these chat features turned on. If they don't, the messages are sent by regular SMS or MMS.

The full set of features in the Google Messages app may be available on only certain devices or carriers. If this full set is important to you, be sure to verify the capabilities before buying a new or refurbished device.

A BRIEF HISTORY OF TEXTING

Back in the dark ages of cellphones, everyone had flip phones (you might remember these creatures) that were nearly impossible to use for texting. It was a laborious task to use the phone's number keys as the alphabetical keyboard. Selecting a letter often required making multiple taps on the same key, as in "Press the 5 key one, two, or three times to enter the letter *J, K, or L, respectively*." That method was called *T9* (text-on-9-keys) texting — and there *had* to be a better way.

In 1993, Nokia came out with the first phone-to-phone texting. For those who wanted more, a company named Danger (as in "Danger, danger, Will Robinson!") was formed in 2000 by three former employees of Apple, and they developed the Hip Top Sidekick. The Sidekick was the first phone to access websites, images, and other data over cellular networks. It was so revolutionary that (Apple cofounder) Steve Wozniak joined the Danger board of directors in 2001.

The QWERTY keyboard on the Sidekick, shown in the image, made typing easy. Revolutionary for its time (back when Paris Hilton was famous for her bedazzled, rhinestone-encrusted Sidekick), it was the cool phone to own, and I used mine constantly. I even upgraded to a newer edition before I bought my first Android.

(continued)

(continued)

This short history is my way of letting you know that the evolution of texting sparked today's mobile revolution. Interestingly, much of the Danger OS was used to develop the first Android phones.

Photo courtesy © Marsha Collier

Danger Sidekick II

Turning on RCS chat features

When you first turn on a new phone and begin using the default Google Messages app, you may be asked whether you want to activate Chat. It's perfectly okay to do so on initial setup, or you can tap Skip and do it later. Sometimes, it's better to acclimate yourself to basic phone and texting features before adding bells and whistles. Don't worry: The OS prompts you again if you forget to revisit any features, or you can access them from the Settings menu.

To activate Chat features at any time, follow these steps:

1. Open Messages and tap on the three vertical dots, usually in the upper right corner, and then tap on Settings.

2. When the Settings menu appears, tap on Chat Features.

3. On the next screen, work through the settings that are available for your handset.

The screen shots in **Figure 10-1** show the differences in chat options between two Samsung phones (both running Android 11). The options are based on the features available for your device and its version of the Android OS.

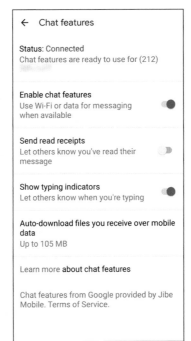

Samsung Galaxy A32 5G   *Samsung Galaxy S10+*

FIGURE 10-1

An important option, available on most phones, is the *multimedia limit* or *Auto-download*, as shown in **Figure 10-2**. You can set a size limit on media files before receiving a warning notice, so pick a reasonable number, based on your smartphone's memory capacity.

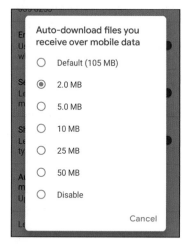

FIGURE 10-2

Some files can be *huge and* take up a lot of space. It never hurts to delete files from your phone after you've backed them up elsewhere.

REMEMBER

Using third-party chat apps

As you may know, you have lots of options when it comes to chat apps. Many of my friends use Messenger from Facebook, Instagram Messages, Telegram, WhatsApp, or other third-party apps. Unless you both use the same one of these apps, you can't share messages with them. To connect on that app, you must go to the Google Play Store, download the app, and agree to an entirely new privacy policy.

For example, I personally feel that Facebook (which also owns Instagram) has way more information about me than it needs, so I try to divert my personal messaging to my phone's text messaging app. Call me paranoid, but I like to keep my data as private as humanly possible in this connected world.

If you chose to bow to peer pressure (I confess that I did, because many friends insisted on using Facebook's Messenger) and you want to use those apps for messaging, caveat emptor. You can deny certain permissions to apps in your phone's Settings.

But that's privacy for your phone only. Know that the promised "end-to-end encryption" only means that no one can access the data as it is being transmitted. A copy of the messages (and everything else you send or receive) from any conversation in the apps remains on that app's servers.

WARNING

When you use apps from a vendor other than Google, you're sharing your data with an entirely new database. Even when you use an app from your phone's manufacturer (such as Samsung), odds are good that you will need to agree to a new privacy policy. You can access security settings and disable certain permissions, but keep in mind that after you alter the settings, the app may not work as expected.

Dress Up Texts in Google Messages

To begin texting, tap to open the Messages app, and tap the mini Messages icon in the lower right. Start to type in the name of the text recipient, and their name should pop-up if they are in your contacts. You can also choose more than one person to send a group text. (You cannot send text messages to landlines.) If a person isn't yet on your list of contacts, tap the icon that looks like a phone keypad and type the number. Now it's time for the magic to happen!

As I note elsewhere in this book, I'm a big fan of emoji and animated files (called GIFs), and I probably add them to my texts more often than an adult should. But you can join the fun in a couple of ways and add a lighthearted touch to your texts: You can easily add emoji from your Gboard (Google Keyboard).

When you look at the Gboard while texting, you find the QWERTY keyboard with all the options mentioned in Chapter 6. For example, **Figure 10-3** shows (on the left) how your keyboard might look with predictive typing. But notice (on the right) that when you're not typing — and thinking of a reply — the top row on the Gboard also has several icons that link to where the fun times begin!

FIGURE 10-3

Each of the tiny icons above the keyboard (as shown in **Figure 10-4**) link to something useful. Tap an icon and, in the window that replaces the keyboard, browse for a special touch you want to add to a text. When you find the right emphasis, tap it and it appears in the Messages app stream above the text message box. Your options for fun features for your text messages are listed in Table 10-1.

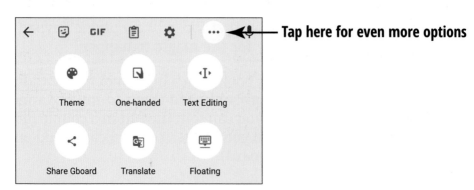

Tap here for even more options

FIGURE 10-4

TABLE 10-1 **Text Message Features**

Icon	What It's Called	What You See When You Tap the Icon
😊	Sticker	The happy face in-a-square leads to groups of cartoon stickers for your texts. You can download more in the Play store.
GIF	GIF	A menu of recently used GIFs, a sideways scrollable menu listing actions or emotions, and a Search icon for typing a descriptive word (or words) to find *just* the right GIF.
📋	Clipboard	Items you may have copied from the Internet or another text message. Find out how to copy text in the section "Manipulate the Text in Your Message."
⚙	Settings cog	Keyboard settings for typing, such as text correction, spell checking, dictionary, and more.
•••	Ellipsis	These three horizontal dots, which mean, in Android, "Tap here for even more options."
🎤	Microphone	A window to your phone's voice-to-text feature where you can speak your message.

REMEMBER

If you get lost in the process of adding fun options (such as GIFs and stickers), you can tap the ABC key or mini-keyboard key that you see at the bottom of the option window. Doing so brings you back to the keyboard and the main typing area for the text you're creating.

TECHNICAL STUFF

Either pronunciation is correct — "gif" (with a hard *g*) or "jif" (with a soft *g*, like the peanut butter) — but the hard *g* is generally accepted. The debate has gone on since 1987, when the technology was invented. The dispute continues, even though the inventor of the GIF prefers the soft *g*.

TIP

For extra fun, you can search for GIFs by celebrity name, movie, or TV show. Hollywood has taken advantage of the popularity of GIFs for PR and often prepares them to promote various properties. Try searching for your favorite show!

Until you tap the arrow to send a text, a small *x, a* minus sign (−), or a similar-looking symbol appears in the upper right corner of the feature you chose. If you change your mind and would rather send a different GIF, sticker, or photo, tap the symbol and the feature disappears. You can then select another or none at all. Your message isn't final until you tap the send arrow.

The Text Message Bar and Emoji

A shortcut to the emoji (small expressive icons I refer to in Chapter 6) usually appears in one of two places: After tapping in the text box, to the right of where your words appear, you may find a small happy (or smiley) face in a circle. Or, that same happy face in a circle may show up as part of the keyboard, to the left of the Space bar.

Tap any smiley, and the image-option icons move to a spot below the keyboard. Tap on the smiley face at the bottom to see a drop-down with a small emoji that you can insert into your texts. Depending on your phone, next to the text box or below those full-color suggested emoji are black-and-white icons you can tap to see a list of other emoji. These represent people; animals and nature; food and beverages; travel and places; activities and events; objects, symbols, and flags; and in some cases, others.

When using emoji, my preference is to tap the Smiley Face in a circle icon at the bottom of the keyboard. When you tap there, a search box opens where you can type a keyword specific to the emoji you're seeking — although browsing emoji often produces unexpected surprises. **Figure 10-5** gives you some ideas for using emoji.

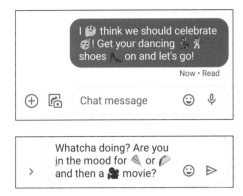

FIGURE 10-5

TIP

The human representations on Android emoji have a generic skin color. By holding your finger on an emoji in the selection box, you can change it to another skin color (as shown in **Figure 10-6**). You can also change the hair color of other emoji.

FIGURE 10-6

Find Even More Texting Options

Tap on the plus sign (+) on the far left side of the text bar to reveal even more fun options for your texting experience. A menu shows up in place of the keyboard; scroll vertically and check out the options, some of which are shown in **Figure 10-7**.

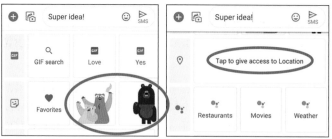

FIGURE 10-7

These additions to your texts have special purposes:

» **GIF:** Another GIF search. Why not?

» **Favorite stickers:** If you've used stickers in the past, the ones you've used most often show up here (as on the left in Figure 10-7) so that you can find them easily.

» **Location:** If you want someone to meet you on a crowded street, or if you're lost, you can tap the location option (shown in the center image of Figure 10-7) and your location on a map will appear. You can view a preview and, if you confirm it, the map appears in the text and sends a link to wherever you are. (Be sure you've enabled the Location icon in the Quick Settings menu).

TIP

Many folks complain about privacy when sharing a location on their phones, but if you ever get lost, you'll be glad you have enabled location sensing in the Quick Settings window shade. I let Google know wherever I am, just in case. You never know when that info will come in handy.

» **Restaurants, Movies, Weather:** You want someone to meet you at a specific location and they don't know how to find it? Google can send them an interactive map! Want to share a movie listing so that both of you can decide? When you're thinking it's warm outdoors and the other person is thinking it's chilly, share the forecast and send them a screen shot!

» **Contacts:** Does your friend need the contact information of someone you know? Tapping the Share a Contact link (shown on the right in Figure 10-7) opens your Contacts folder. Search for the person whose contact information you want to share. Tap the person's name, and a clickable mini file with all the contact info appears in the text box.

WARNING

If you have inserted private notes in your contacts, this information will transfer with the contact file. Instead of using the Share a Contact option, I recommend that you copy and paste the needed information from your contacts and send that data as a text by itself.

» **Attach a File:** Need to send a file? Here you can send a PDF (where you may have saved a recipe or an article from the web). When you tap the Attach File option, you're directed to an area where your files are stored. Just select and send! (My husband may find it annoying, but I send PDFs of Internet articles to him all the time.)

Share Photos and Videos in Texts

Sharing photos and videos via text messaging is the easiest task of all. Tap the Photo icon to the left of the text message box and you see the last few photos you've taken on your phone. At the end of the limited preview images, you can tap to open the Gallery (see **Figure 10-8**). From the Gallery, tap a photo to add it to the message.

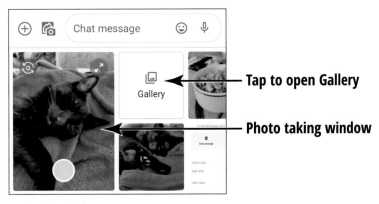

FIGURE 10-8

Also, a picture taking window opens (or, on certain phones, you can see the window by tapping the Camera icon next to the Photo icon). If the window is too small to work with, just tap on the circle with two arrows in the upper right to fill the screen. To take a selfie (picture of yourself), tap the Camera Mode icon (a circle with arrows) on the left, and the camera switches to Selfie mode.

REMEMBER

Attaching a picture to a text message is a great idea, but taking the picture from the camera in the Messages app — which isn't as robust as your phone's Camera app — may not be so great. I recommend that you take a photo with your phone's Camera app and attach it from the Gallery.

TIP

Saving a photo you've received in a text is as easy as attaching one to a text you send. Tap the photo in the message to make it full screen size. A down arrow appears at the bottom of the screen; tap there, and the photo is saved to your Messages app's Stored Media Folder in the Gallery.

Voice-Type (Dictate)

It took me a while to cotton to the idea of talking to my phone. I'm still not comfortable dictating to it when others are in the room. But (big *but* here) dictating to your phone (as I describe in Chapter 6) becomes more accurate with regular use — and certainly speeds up your communication. (I can speak more swiftly than I type on Gboard.)

You can find two Microphone icons: one in the text box and one on the options bar of the keyboard. Here's what they do:

>> On the far right end of the text box (to the right of the Emoji Browse icon, shown in **Figure 10-9**) is the Microphone icon (on some phones, audio lines), which enables you to record voice messages. Tap it and talk. If you make a mistake, tap the tiny *x* in the corner of the recorded message to delete it. You can also type additional text to accompany your voice message.

FIGURE 10-9

» Tap in the text box and another Microphone icon appears, but this time on the texting options bar above the keyboard. Tapping this icon enables you to dictate words that are translated and then show up in the text message box. **Figure 10-10** illustrates steps in this process.

FIGURE 10-10

REMEMBER

I usually dictate a message and then go back and use the keyboard to edit the translation. This rechecking is important because — trust me — dictation still isn't 100 percent accurate.

Manipulate the Text in Your Message

After you type out your missive, you may want to reread it for accuracy. I know that if I hit Send immediately, my message will no doubt feature a spelling error (which I then need to quickly correct in a follow-up text). I have made some doozies!

When you want to edit, here's how to do it:

- » **Delete text (or emoji).** Tap to place the cursor to the right of the character(s) you want to remove and tap the backward arrow (Backspace key just above the Return key) to remove the offending text or emoji. You can tap more than once to delete more than one character.

- » **Add text.** Tap anywhere within the text message and a cursor appears. Type the words or emoji you want to add, and press Send.

- » **Work with specific or multiple words.** Tap on a word and hold briefly within a text message you're writing. The word becomes highlighted (it's light blue in **Figure 10-11**), and handles appear at the beginning and end of the word. If you want to work with more than the one word, press and drag a handle to highlight those words.

 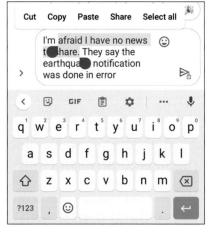

FIGURE 10-11

A pop-up menu with editing options appears when you highlight text in your message. Although the options vary among messaging apps (and sometimes change because of the content of the text you highlight), you find options such as

- *Cut:* Remove the text from the message and place it on the clipboard — in case you need it elsewhere.

- *Copy:* Yep, you can make a copy of the text and place it on the clipboard.

- *Paste:* Paste some text that you previously copied from the clipboard into the message.

- *Undo:* Deletes the highlighted text from the message.

Tapping the three vertical dots on the right side of the pop-up menu (depending on the messaging app) enables you to select all text in the text box, paste copied text as plain text, see the clipboard, or perhaps even share the message.

Schedule a Text Message for Later Delivery

I often think of things to tell people as I'm just going to bed. This isn't the best time to send a text message, because many people don't set their phones to Do Not Disturb, and an inconsequential message triggers a notification sound and wakes them up.

Google Messages to the rescue! When you've finished writing the text, long-press on the Send arrow. A menu, as shown in **Figure 10-12**, pops up. Just select the time you want the message to be delivered, and Google Messages arranges everything once you tap the send arrow again. This setting has never failed me.

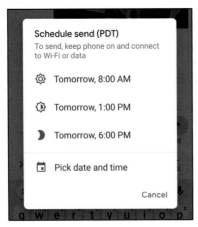

FIGURE 10-12

Act on Text Messages You Receive

There's a bit of etiquette to everything in life, and many feel that *private messages* should remain *private* unless you are asked to share them. For example, you may notice that people ask you to send a private message (PM) on social media so that (guess what?) it stays private. If you must share some tidbit from the message, copying and pasting the non-private information to share can do the trick for that.

Despite the etiquette admonishment, you *can* forward a text to another contact or share it on a social media platform. You can even *Like* it and comment with an emoji (but only when you're using RCS chat).

To perform an action on a text message you've received, long press to select it and two menus appear: one at the top of the text screen, and another with animated emoji over the text. The top menu (on the left in **Figure 10-13**) enables you to copy the text to place elsewhere, delete it (by tapping the garbage can), or make it a favorite by tapping the star. Tap the three vertical dots for extra options (shown on the right in Figure 10-13).

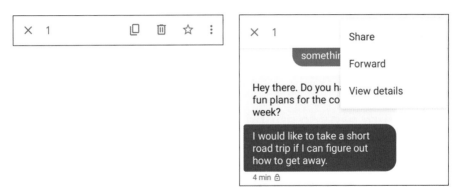

FIGURE 10-13

Above the text you selected, the pop-up menu (see **Figure 10-14**) shows thumbs up, smiling face with heart-eyes, face with tears of joy, wow, tear, angry, and thumbs down emoji. Tap the emoji that connects with your feelings, and you (and the recipient) see it as a small emoji in the corner of the text message.

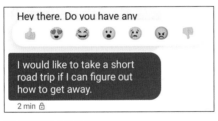

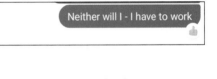

FIGURE 10-14

Share and Print Documents, Email Messages, and Web Pages

No one explains how to share or print documents, email messages, or web pages in any of the books I've read, so I give you the 411 here. When you want to print a copy of a document — perhaps an email with a QR code or a page on the Internet — you can find a way to do it.

Printing from Gmail

First, you need to connect your phone to the same Wi-Fi network as your printer. Some apps upload documents to the cloud and send them to a wireless printer, but — really? You don't need to sign up with a third party to print documents. Follow these steps to print an email message:

1. Open the email in the Gmail app (or click to print a web page) and tap the three vertical dots in the upper right corner. A drop-down menu appears, giving you the option to print the email.

2. Tap Print. On the standard printing screen that appears, you see the image you're printing, and you can choose a printer at the top (if it isn't showing already) and set printing options (such as number of pages and others).

3. When everything is ready, tap Print and your email flies through the air to your wireless printer.

4. You see a screen where you can tap the Printer icon to begin the print process.

Printing from a web page

You have a couple of ways (of course) to print and share. If you've arrived at a web page by way of Google Search, you see the Share icon (highlighted in **Figure 10-15**) at the top of the page, and you may be able to print that way.

FIGURE 10-15

To see more of the options spelled out in the following list, you need to tap the three vertical dots and choose Open in Chrome (or Open in Browser) from the menu options. Tap the three vertical dots again and tap Share from that menu. A panel appears at the bottom of your screen with one or more rows of scrolling options, including Print. Tap Print and follow the prompts from there.

You have other options for sharing available from the panel. (You may need to scroll left through the options to reach the More option and tap it.) Some of the options other than printing are

» **(Take a) Screenshot:** Tap here and your phone takes a photo image of the page.

» **Copy Link:** Tapping here gives you the URL link for the page so that you can add it to a text or email message or post it on a social platform.

» **Send to Your Devices or Nearby Share:** If you want that link to appear on your laptop or on another Android device, tap here to bring up a list of available devices. You will have to turn on Bluetooth on both devices, then tap to notify the other device that you have a link in transit. Accept it on your other device, and then (like magic) it opens in your browser.

» **(Make a) QR Code:** It's all advanced stuff here — impress your friends with your technical skills by generating a QR code, something similar to the example shown in **Figure 10-16**. Tapping the QR code option creates a QR code that can be read by another smartphone and then opened on that device.

You can click to download the created code and send it as an attachment or just hand your phone to the recipient so that they can scan it directly from your phone. They do so by tapping the Google Lens icon in their camera, or by selecting the Read (Scan) QR Code icon from the Quick Settings menu.

FIGURE 10-16

Chapter **11**

Managing Email with the Gmail App

started using Gmail in 2010 (after being sure it wasn't a flash in the pan). It was new to me and seemed to be the most advanced email system. Besides, it was cool (in those days) to have a Gmail email address. It seems that I bet on the right horse, because in 2020, the platform had 1.08 billion users and 53 percent of the United States email market.

My Gmail mailbox has become my personal file box, containing a decade of saved emails because Gmail allows you to set up folders for different projects. I can find every email from my daughter, my lawyer, my editors, or whoever else sends me a message — merely by searching. Obviously, keeping all these messages takes up space on Google, but you are given an initial 15 gigabytes (GB) of free storage. (Chapter 1 gives you the details on paying a small amount for a bit more space.)

In this chapter, I tell you about the features of Gmail that make it an excellent choice for an email app. I also show you how to use Gmail for sending and receiving email messages, organizing and archiving important messages, and even printing messages.

Discover Gmail Features

One reason for the popularity of Gmail is its use of filters that detect spam and malicious email. The filters are powered by machine learning, and the algorithms do a terrific job of ferreting out spam. When the spam email comes in, the message is deposited directly in my Spam (or Junk) folder, which is emptied automatically every 30 days.

REMEMBER

Spam is not only annoying — it also clutters up any email box and can cause you to lose track of vital email.

If your phone's manufacturer did not include it, you'll have to download the Gmail app from the Google Play Store and sign in to your Google account. After the Gmail app installs, you can tap your initial or your photo (in the upper right corner) to access your Google account. (If you've used Google for a while, it knows who you are.)

Revealing the Gmail app's main menu

Gmail offers a wealth of preset folders, with quick access to them from the main menu. The *main menu* on the mobile app, which is scrollable, provides access to other Google apps (such as Calendar and Contacts) and to Gmail settings.

In the Android Gmail mobile app, you don't see a folder on the main screen unless a new email message has arrived and is displayed in bold type. To take a look at all Gmail folders (as well as any you have personally set up with filters) available from the main menu (as shown in **Figure 11-1**), open the Gmail app and tap the three horizontal lines (affectionately known as the *Hamburger menu*) in the upper left corner of the screen.

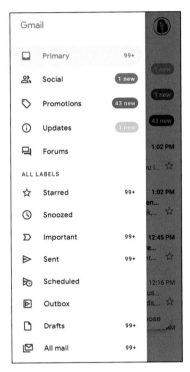

FIGURE 11-1

Tending to mailbox organization

I want to tell you a bit about Gmail's folders, which on the computer desktop version of Gmail, appear as divider tabs across the top of the email list view. When you open Gmail, you may not see all of the folders (described in the next section "Scoping out the main mailbox"); if that's the case, you need to activate the folders.

You can activate folders on the Gmail desktop version or on your phone. On your phone, tap to open Gmail and follow these steps:

1. Tap the Hamburger menu icon on the upper left to see the options menu. Scroll toward the bottom of the menu to find a link for Settings.

2. Tap Settings, and then tap your email address. The resulting screen shows many options for your email account settings in the app.

3. Click the Inbox Categories option. In the resulting screen, I strongly advise you to activate folders (refer to Figure 11-1) by tapping the check box for the category name (Social, Promotions, Updates, and Forums, for example).

REMEMBER

Activating mailbox folders in Gmail may be a minor bit of organization, but it can massively improve your email experience.

Scoping out the main mailbox

A handy feature of Gmail is that the main mailbox is divided into multiple sections, or *folders*:

TIP

» The **Primary** inbox folder holds the most important email you receive from people (versus blanket emails from businesses) and those you select to appear there.

If an email appears in a place other than the Primary inbox folder (for most important emails, according to Google's algorithms) — and you want to move it there — just click the Star symbol next to the email. After you click the star, that sender gains priority and their email messages always move directly to the Primary inbox folder, with less chance of being lost.

» The **Social** folder holds social media notifications from sites like LinkedIn, Facebook, Twitter, Instagram, and others. Also, it's the primary entry for news aggregation site emails. (I move those to the Primary folder.) It doesn't sound like this would be a lot of email, but it's helpful to have social media segmented from my work messages and personal family messages.

» The **Promotions** folder is for ads from companies you've done business with or with whom you've signed up for notifications.

» The **Updates** folder holds emails that are marketing oriented, such as Amazon order information, travel, bills, and paid bill receipts. When an airline, hotel, or event confirmation shows up in Updates, I click the star so those emails will show up in my Primary folder.

If you also use Google calendar, official incoming travel confirmations are automatically added to your calendar. Very handy when it comes to organizing travel information!

» The **Forums** folder contains emails from news lists you've signed up for.

Send a Gmail Email

Although I'm pretty sure that using an email program isn't a mystery to you, please allow me to show you how it's done in the Gmail app. I'll wager that sending a message in Gmail is the easiest task you will learn today. Follow these steps:

1. Tap the Gmail icon to open the app. (***Note:*** Putting a Gmail shortcut on the Home screen adds convenience for this step. See Chapter 5 for instructions.)

2. On the resulting screen, tap the small Pencil icon at the bottom, next to the word *Compose*. The next screen shows you the email form.

3. Starting at the top of the form, type the name of the intended recipient (or their email address) on the To line. As you type, Gmail suggests names from your contacts that match the characters you're typing. If you see the recipient's name in Gmail's suggestions, tap the name to place that person's contact information on the To line.

You might want to compose the email (see Step 5) *before* inserting the recipients' names. This is my policy because I've accidentally tapped the send arrow prematurely on an email that wasn't properly edited. I faced some embarrassing consequences.

You can add other recipients' names to the To line after you complete adding the first recipient. Just start typing the next recipient's name (or email address) at the flashing cursor.

4. If you want other people to see this same email, but not as the official To recipient, tap the small down arrow to the right of the form's To line. Two new lines (below *To*) appear (see **Figure 11-2**) and offer these options:

- **Cc (*carbon copy*):** When you add names to the Cc line, everyone on the To and Cc lines receives the same email and can see the names (and email addresses) of all recipients.

- **Bcc (*blind carbon copy*):** This option is sneakier. When you type a name or an email address on the Bcc line, the official To recipient of the email can't see that you copied the email to a Bcc party. Using Bcc also protects those who don't want to share their private email addresses.

FIGURE 11-2

5. Type a subject on the Subject line, and then type the text of the email message where you see the words *Compose email.*

6. After you proofread the email, tap the small arrow in the upper right corner to send it on its way.

GMAIL SENDING OPTIONS

If you write emails late at night or are awaiting further information that needs to go in the email, Gmail has options that can help. Before clicking the send arrow, tap the three dots next to the arrow to see a special menu that applies to the email you're working with.

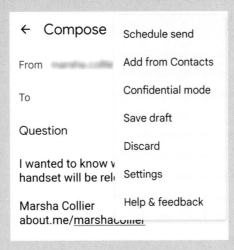

Here are the most important options from this menu:

- **Schedule send**. Tap this option, and you can send the message at a later date and time.

- **Confidential mode**. Tap this option to choose an expiration date for the message or revoke access to it at any time. The recipients' ability to forward, copy, print, and download the message is disabled. (But they could take a screenshot of it.)

- **Save draft**. Tap this option to save the email in a Drafts folder for sending at a later date.

- **Discard**. Tap this menu option to abandon the message; it's your delete button.

Adding an email signature

Having a preset signature show up on every email makes sending them out quicker. In addition to your name, you can add your mobile phone number, your business name, or a link to an Internet page (refer to Figure 11-2). To preset a signature, follow these steps:

1. Tap the Gmail icon to open the app.

2. Tap the Hamburger menu (the three horizontal lines) in the upper left corner.

3. Scroll to the bottom of the resulting screen and tap Settings.

4. Tap your email address on this screen.

5. Scroll the next screen a little and tap the Mobile Signature option.

6. When prompted, type the text you want to appear as your signature on every outgoing email.

7. Tap OK.

Creating a Vacation Responder email

In the business world, most people send a standard autoresponse email whenever they go on vacation (or are taking a break from email). It's called an *out-of-office* (or *OOO*) email response.

I find few things more frustrating than sending an email to a friend and getting no response. And so, I craft OOO email messages for use in my personal email account. (I know some friends, if I didn't respond, who'd definitely fear that the worst had happened.)

Follow these steps to set up an autoresponse:

1. Tap the Gmail icon to open the app on your smartphone.

2. Tap the Hamburger menu (the three horizontal lines) in the upper left corner.

3. Scroll toward the bottom of the resulting screen and tap Settings.

4. Tap your email address to open the screen with settings to customize it.

5. Tap the Vacation Responder link, shown on the left in **Figure 11-3**.

6. On the next screen, toggle the responder to On, which activates a form where you can add a vacation responder email message.

7. Set the dates (first and last) that you plan to be away and want to use the autoresponse email. Then type the relevant info into the email form.

8. If you want only your contacts to receive this email, tap to put a check mark in the Send Only to My Contacts check box.

9. When you're satisfied with the dates and your email message, tap Done.

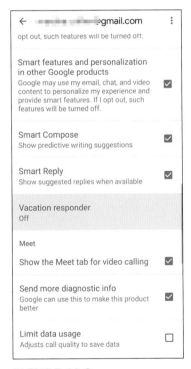

FIGURE 11-3

From now on, anyone who emails you during the period you set receives the vacation responder message you wrote. Check out the nearby sidebar, "My viral autoresponder email," for an example.

MY VIRAL AUTORESPONDER EMAIL

When my husband and I got married, we went on a prolonged honeymoon. The autoresponse email I crafted went a little viral and is quoted in several posts on the Internet. I tried to be businesslike (just in case an email came from a business connection) but needed to keep the message personal.

I share the following email with you for inspiration. I hope it gives you an idea of how to craft your own masterpiece:

> I am currently out of the office for my wedding and honeymoon.
>
> I know I'm supposed to say that I'll have limited access to email and won't be able to respond until I return — but you know that's only partly true. My devices will be with me, and I can respond if I need to. And I recognize that I may need to interrupt my honeymoon from time to time to deal with something urgent.
>
> That said, I promised my husband that after the wedding I am going to try to disconnect, get away and enjoy our honeymoon. So, I'm going to experiment with something new. I'm going to leave the decision in your hands:
>
> If your email truly is urgent and you need a response while I'm on my honeymoon, please resend it to ********** and I'll try to respond to it promptly.
>
> If you think someone else at The Collier Company might be able to help you, feel free to email my assistant at ********* and she'll try to point you in the right direction.
>
> Otherwise, I'll respond when I return. . .
>
> Warm regards,
>
> Marsha

Perform Basic Gmail tasks

In the earlier section "Send a Gmail Email," I show you how to send an email (and sneak in a Bcc). This section goes over some other basics that can enhance your Gmail experience.

A table of common email tasks

Table 11-1 presents common email tasks and how to accomplish them.

TABLE 11-1: **Email Tasks and How You Do Them**

The Task	How to Do It
Open	Tap a new email (which appears in bold) from your email inbox list, and a new screen opens to show the full email message in a scrollable window.
Close	Tap the back arrow in the upper left corner (of an open email) to return to the inbox list. The closed email remains in your email inbox list but no longer appears in bold.
Delete	Tap the little Trash Can icon at the top of the open email to delete just that one message. You can also delete one message or multiple messages while looking at the list of messages on the Inbox screen. Tap and hold the email you want to delete, and a check mark then appears on the left side of the email, to show that it's selected. After you select the email, or emails, that you want to delete, tap the Trash Can icon at the top.
Archive	Swipe right on an email from the inbox list view to archive it. Archiving removes emails from the list in the inbox but keeps them in your Gmail. If someone responds to that archived email, it shows up again in the inbox list. (I'm a rebel. I don't archive my emails.)
Mark As Unread	From an open email, tap the small Envelope icon at the top of the screen, or tap and hold to select the email from the inbox list view and then tap the Envelope icon at the top of the screen.
Move	Tap the three vertical dots in the upper right corner of an open email to open the Options menu. Tap the Move To option, and a screen showing a list of other mail folders appears. Tap the folder name to move the email to a folder in the list. The folder options presented vary, depending on the folder where the email you want to move resides now.
Search	Tap the Search in Mail box at the top of the inbox list view. On the resulting screen, type a search term where prompted. Then tap the Search icon on the keyboard or the back arrow in the upper left corner. You can find emails from a specific person or an email that contains specific text.

After reading an email in the Gmail app, you can easily delete it by tapping the tiny Trash Can icon at the top of the screen. *Note:* When you tap the trash can, you see a banner appear briefly at the bottom of your screen giving you the chance to Undo your deletion action. Anything you delete is stored in the app's Trash folder and can be restored up to 30 days after you delete it — either from your phone or from Gmail on your PC.

Printing an email from Gmail

To print an email, tap to open it and then tap the Options menu (the three vertical dots) in the upper right corner of the screen. Then follow these steps:

1. Tap the Print option on the menu, as shown in **Figure 11-4**.

FIGURE 11-4

2. A new screen opens, showing you the entire email.

3. At the top right of the new screen, tap the down arrow to select whether you want to print to a Portable Document File (PDF, which will save in your phone's My Files (or File Manager) apps, or to one of the printers on your home Wi-Fi network.

4. For even more control of printing, tap the down arrow at the top of the print area to expose a menu where you can define which pages to print, the paper's size and orientation (Portrait or Landscape), the number of copies, and other options.

Link to Other Apps and Gmail Settings

At the bottom of the folder list that you access from the Hamburger menu (the three horizontal lines), you can see links to other Google apps; Calendar and Contacts are two popular ones. You also can see a link to the Settings app — the settings are worth taking a look at. By using these settings, you can manage notifications, change how you view messages, set defaults, and choose many other settings. For example, **Figure 11-5** shows where I changed the default swipe action so that I swipe right on emails to delete them.

Feel free to poke around in the Gmail settings. You won't break anything, and you may just find a way to improve your Gmail experience. If all else fails, at the bottom of the list is a link to the Help and Feedback area, which can help you solve your Gmail problems.

I love using Gmail, and don't think I'll ever be convinced to switch.

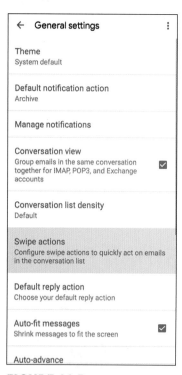

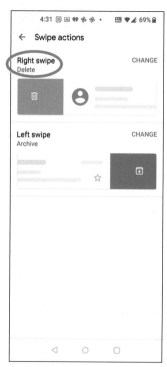

FIGURE 11-5

IN THIS CHAPTER

» **Understanding smartphone cameras**

» **Matching a camera with your needs**

» **Taking a look at the options**

» **Using the Pro camera**

» **Editing your masterpieces**

Chapter **12**

Choosing and Using a Smartphone Camera

R emember when you had to cart your camera film all the way to the drugstore (or Fotomat) to have those folks develop a roll of photos for you? For a roll of film, the processing cost was about $2 and each print (whether viable or not) cost 30 cents. Given these costs, you had to be *selective* about the number of photos you shot and their subject matter.

Today, photos are basically free. Digital photos can reside on phones or laptops, and in the cloud. From the digital versions, you can make beautiful prints for around 25 cents, and booklets from Google Photos starting at $9.99. I've given booklets from Google Photos as holiday gifts — and people love them. (Go to `google.com/photos/printing/photo-books` for more information.)

In any case, your photos are only as good as your photo talents and your camera. Today's smartphone cameras can enable amateur photographers to produce Ansel Adams-quality pictures.

In this chapter, I cover the essential information — specifications, features, and functions — that you need in order to produce great-looking photos from the camera on your smartphone.

Examine Smartphone Cameras and Brands

For the majority of us humans, shooting photos with smartphones may be the most satisfying photographic experiences of our lives. The tiny cameras in smartphones are technological wonders that can make artwork out of a simple sunset. The ultimate application for your photos probably depends on what you're taking photos of. (I have one friend who texts me photos of her toddler grandchild riding around on a Roomba).

Paying the right amount of attention to reviews

If you're looking for a new Android smartphone, reading reviews — including what is said about the phone's camera — isn't a bad idea. Just know that online reviews are usually posted within the first 24 to 48 hours after a reviewer gets their hands on the smartphone. Rarely do these reviews delve deep enough for you to form a complete picture of the quality or versatility or flaws of the built-in camera.

TIP

If you're buying a smartphone *specifically for the camera,* check out reviews, yes. Then go to the manufacturer's website and read about the camera's specifications. Compare the camera's specs with others on gsmarena.com.

The proof of the camera's suitability is in the phone itself. Go into a retail location to take a look at the camera features and photos with your own eyes. Match what you want from a camera phone with what you see as well as the sales pitch you hear.

Phones with camera brand collaborations

Famous camera brands that you may recognize have made partnerships with today's Android phone manufacturers to produce outstanding cameras in their flagship phones. Some smartphone brands even partner on the development of new technologies with the camera companies; others may only license the name. This list describes some examples:

» **OnePlus:** OnePlus just started a 3-year, $160 million partnership with the Swedish camera brand Hasselblad. Its goal is to codevelop its next-generation phone cameras. By engaging in extensive research and development (R&D) together, OnePlus and Hasselblad plan to improve existing color tuning and sensor calibration features. Their latest camera update incorporates Hasselblad's famous XPan Mode (from the analog days) into the OnePlus 9 series.

» **Huawei:** Although new models are not available in the United States right now, for the past few years the Huawei phones have been respected as having some of the finest mobile cameras in the world. The 2019 P30 Pro is still being reviewed as one of the best smartphone cameras ever. Huawei partnered with Leica to develop its entire camera module. To this day, I keep my P30 Pro in my purse to use as just a camera.

REMEMBER

Google's **Pixel** phones are recognized for shooting some of the best-looking photos — without a camera company collaboration (probably due to computing power since they're making their own chips now). **Samsung** phones take great-looking photos as well. I've heard that Samsung may be partnering up with Olympus for their Galaxy S22 Ultra. Who knows what the future holds?

Take a Camera-Spec Safari

The best features to look for in a phone camera are based primarily on what you want the camera to take photos of. I'm a generalist, and I want to be able to take selfies *and* every other type of photo imaginable. I talk about essential features in this chapter's later section "Exercise Your Android Camera's Capabilities," but for right now, let's dispel some myths.

Enough megapixel, but not too much

People seem to want more and more megapixels (MP) in their cameras. (A *megapixel* is a million pixels.) But long story short, you don't need as many as you think. A higher megapixel specification in your camera produces a larger photo file size. Because photos are stored on your camera, this larger file size can adversely affect its storage space. But adding pixels means adding density to the camera's sensor. With more density, you can successfully zoom in for photos taken far away from the subject.

TECHNICAL
STUFF

When you look at your television, are you happy with the high definition (HD) picture? Full HD has only 1920 x 1080 pixels, which means that a high definition TV picture has only enough pixels to approximately equal 2MP (2,073,600). A 4K (K stands for *kilo* or *thousand*) resolution display increases the number of pixels to equal 8MP. So why do you need 40 MP in your phone's photos?

I am not a professional photographer, by any means. But I'm delighted when regular folks (like me) can capture great-looking photos with their phone's camera. **Figure 12-1** shows a 6MP photo I shot in Paris with a 2018 Huawei Mate 20 Pro. I wanted to capture the Eiffel Tower at night.

The photo on the figure's left shows the full picture as I took it that night. The version on the right shows the detail area captured when I expanded (pinched outward) on the photo and took a screen shot. The detail area reveals the statue of a horse at the base of the tower. More detail is what using more megapixels can provide. When I was shooting the photo (from a moving boat), I pinched inward to squeeze more details into the frame.

Photos courtesy of © Marsha Collier

FIGURE 12-1

Home in on subjects with zoom

Figure 12-2 shows a 2020 Rose Parade float, which is beautiful from a distance. Did you ever wonder what the floral details look like? The top photo in the figure shows a 9MP photo (pinched inward to decrease the zoom) from a Samsung Galaxy Note 10+.

Zoom (pinch) outward and (thanks to more megapixels) you can cut out parts of a photo that lack the detail you're after. You still have a clear photo with an entirely new focus, as shown on the bottom of Figure 12-2.

Photos courtesy of © Marsha Collier

FIGURE 12-2

Another marketing myth is about how the camera accomplishes its zoom. There are two options (maybe three – which combines both) for zoom:

» **Optical:** The type of zoom that most closely replicates the high-quality cameras of the past. This type of zooming (or zooming out) is accomplished by lenses, which is why the newest cameras have more than one lens on the back.

» **Digital:** Where the high-tech comes in. Software in the camera digitizes the picture and zooms by way of in-camera processing. In the olden days of digital cameras, this meant getting fuzzy pictures because the pixels were approximated by early-stage software. But now smartphone cameras have powerful image processors, artificial intelligence (AI), and fancy algorithms to enlarge pixels in the center of the photo and crop out the rest.

TIP

Pinching out to use the **Super Macro** mode on your camera is the way that you can capture those magical ultra-close-up pictures. Your camera can record vivid details that even your eye can't see on its own.

It isn't only the most expensive phones that have zoom features like these. The 2020 Samsung Galaxy A32 5G has four cameras (and a flash) on the back; you can purchase a new one for about $250. If you're looking for the best camera possible, consider buying a gently used or certified, preowned older phone.

Here's the moral of this safari story: If your main goal is to take great-looking photographs, you have a little research to do. Also don't forget, "Practice, man, practice."

Exercise Your Android Camera's Capabilities

Remember the Kodak Instamatic? You may have even used one (I did). It was a simple little box camera where you looked through a lens. It made a "click" noise when you pressed the Shutter button, and it took a photo. A simple process. Granted, the final prints weren't good — but you had a picture.

Surprisingly, today's smartphone cameras can be as simple as the Instamatic. You don't need to learn to use all the bells and whistles to take a wall-hanging-worthy photo. You can use digital photos from even a 5MP phone camera to enlarge and print on canvas. I have a couple of such photos on canvas hanging in my house right now.

The point here is, considering the quality of the smartphone cameras today, you really don't need to know much. Nor does it cost you a penny to shoot a photo — and you can delete them as quickly as you shoot them.

Just point-and-shoot either stills or video

When you ask people (usually, iPhone owners) which phone cameras take the best photos and videos, you generally hear "iPhone." (But recognize that many people use "iPhone" as a generic phrase for a smartphone — kinda like a Kleenex.) That's simply no longer the truth. Android smartphone cameras frequently earn top rankings in digital photography.

To take a museum-quality photo with your Android phone, follow these steps:

1. Tap the Camera icon on the Home screen or in the app drawer to open the app.

2. Wait a second for the camera to focus, pinch inward or outward to frame the subject, and tap the Shutter button.

   Museum quality, of course, assumes that you have composed a beautiful image.

REMEMBER

You can record a video just as easily as you take a photo, by following these same steps. Except, before Step 2, choose the Video option you see on the scrolling menu at the bottom of the camera's image viewer (shown in **Figure 12-3**). Tap the Shutter button to start recording; tap again to stop.

TIP

You can turn the camera lengthwise for a photo in Landscape mode and hold it straight up for Portrait mode. The phone adjusts your screen to the proper viewfinder mode.

Add interest with your camera's tools

In the preceding section, I show how easily you can capture a simple digital photo that still retains quality. Keep in mind that Android cameras come in many styles and formats, but the basic tools of these cameras are quite similar. In this section, I go over a few of the basic tools.

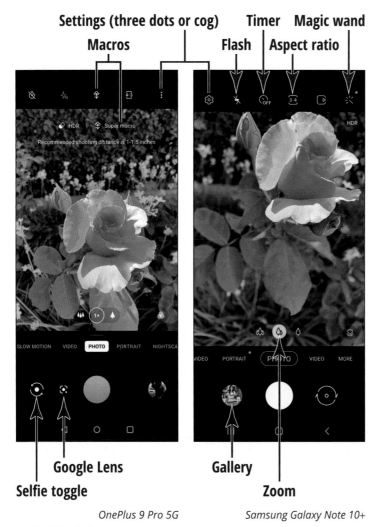

Settings (three dots or cog) Timer Magic wand

Macros Flash Aspect ratio

HDR Super macro

Recommended shooting distance is 1-1.5 inches

SLOW MOTION VIDEO **PHOTO** PORTRAIT NIGHTSCA

VIDEO PORTRAIT PHOTO VIDEO MORE

Google Lens Gallery

Selfie toggle Zoom

OnePlus 9 Pro 5G *Samsung Galaxy Note 10+*

FIGURE 12-3

Refer to Figure 12-3 for screens from two brands of phones with their camera-control icons called out. You find similar icons on many brands of phone cameras. Table 12-1 identifies the icons and what they do.

TIP

On Pixel cameras, you have to swipe downward on a small triangle (or arrow) to find some of the tools listed below.

TABLE 12-1   Basic Smartphone Camera Tools

What It's Called	What It Does
Selfie toggle	Switches your camera between front-facing and back-facing views. To take a selfie of yourself and friends, toggle this button to use the front-facing camera.
Google Lens	Supplies an image recognition technology developed by Google. See the section "Expand your reach with Google Lens" for how it integrates with your camera.
Timer	Sets the camera to shoot a photo after the period, in seconds, that you specify. To take a group picture with you in it, set your phone on a rock, a tripod, or any stable surface. Then set the timer and move into the picture.
Flash	Turns on your camera's flash, which, odds are, you'll never need. Your phone's camera is capable of taking pictures in even the lowest light conditions. (I haven't used a flash in any of my pictures in over five years.)
Macros	Sets your camera to take a close-up photo. When you move in close on a subject, this mode often turns on automatically.
Aspect ratio	Shows the ratio (width to height) of an image. Tap the tool and adjust the default by tapping another selection.
Settings cog (or three dots)	Opens to even more options for your camera and its picture-taking modes. In Settings, you can personalize the defaults for your camera.
Magic wand	With a tap, adds filters or special effects to an image. When you have the time, take a look at how the various filters can alter an image.
Zoom	Zooms in or zooms out on the subject with a tap or a swipe.
Gallery	Shows the photos you've taken. Tap here and compare multiple photos or see whether you need to reshoot the image.

Expand your reach with Google Lens

Tap the Google Lens icon to open a surprising tool: Its seemingly futuristic artificial intelligence (AI) draws from the massive data-bases in Google servers to identify items in the image you're viewing with your camera. **Figure 12-4** shows how Google Lens helps me identify a *Plumeria alba* plant from the flower framed on a Pixel 3XL.

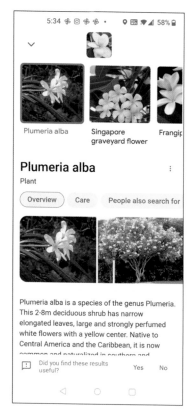

FIGURE 12-4

You can use Google Lens in many ways to connect real-world items (in photos) to data such as restaurant and book reviews. It can copy data, for example, by making a contact from a business card, or add events to Google Calendar by *scanning* (copying text) from books, PDFs, or flyers. It can even translate text, and the results will display right on your screen. If you scan pages of text, you can copy and paste the text into an email, note, or a Google Doc. It's not a perfect recognition tool, but it's getting better all the time.

You may have to download the Google Lens app from the Google Play Store if it's not included in your camera app. On some phones, you can also use it in Google Assistant. When you say "OK, Google," it pops up on the screen, as shown in **Figure 12-5**.

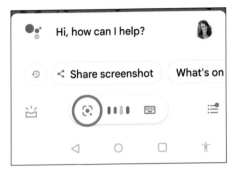

FIGURE 12-5

Access camera features on the scrolling menu

At the bottom of the camera's image viewer, a horizontal scrolling menu enables you to select a photography feature or function for a photo. Each option listed on the menu offers more ways for you to be creative with your photos.

The menu options vary from brand to brand, but generally you find the options described in this list:

» **Photo:** Take a picture, of course.

» **Video:** Film a video.

» **Nightscape:** Make interesting effects when shooting photos at night — for cityscapes, especially.

» **Portrait:** Add special filters to make a photo of someone's face especially flattering.

» **Panorama:** Take a photo in Landscape mode with a wide-aspect ratio.

» **Slow Motion:** Record video at a fast frame rate. When you play the video, the movements appear slower.

» **Pro:** Replicate the professional controls of a fancy single-lens reflex (SLR) camera on your phone's camera. I devote the next section to a brief discussion of Pro mode.

The options may be endless, depending on the brand of camera your phone uses. The best part is that you can try them all out on different subjects, learn the modes and tools that you personally like, and delete less-than-stellar photos and videos.

TIP

To find even more advanced camera settings and options, you can swipe up, or on Samsung camera apps, tap More from the app's scrolling menu. **Figure 12-6** shows the options on a One-Plus (left) and a Samsung (right) smartphone camera.

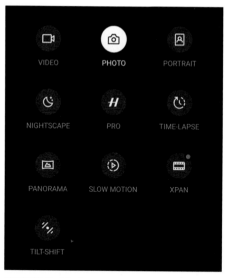 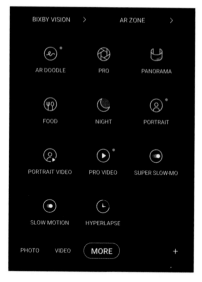

FIGURE 12-6

Go Pro with Pro Mode

When you use *Pro mode* — which is for professional or superhobbyist photographers — each setting can make a difference in the quality and personality of your photos. The camera's automatic functioning is disabled in Pro mode, and the quality of the photo is all up to you. If you've ever enjoyed the creative rush of photography while using an older-model single lens reflex (SLR) camera, here's where you can re-create that creativity but see the results immediately.

Discovering the Pro settings

Here are three universal settings that appear in Pro mode for all phone cameras:

» **White balance:** Affects the quality of the photo because it affects its color tones. A neutral light color occurs in the temperature range from 5500 K (degrees Kelvin) to 5700 K, and you can move up or down on this scale to achieve a warmer or cooler tone, respectively.

As a frame of reference for white balance, sunlight measures about 5900 K, an old-fashioned incandescent bulb ranges from 2700 K to 3000 K, and a fluorescent light or a blue sky measures from about 8000 K to 12000 K.

» **ISO:** Shows the sensitivity (to light) of the camera's sensor. If you see a nighttime photo with a lot of visual noise, you can bet that the noise would decrease if the picture were taken at a higher ISO. The higher the ISO, the better for low-light conditions.

For those who remember buying film, the tricky ISO setting corresponds to the film speed. Daylight film was ISO 100, and Tri-X black-and-white, low-light film was ISO 400.

» **F-stop/aperture:** Refers to the size of the lens opening (*aperture*) and its ratio to the lens' focal length (*f-stop*). Although f-stop and aperture are not *exactly* the same thing, many newer cameras — like the one in your smartphone — combine these into a single setting for ease. Together, they control how much light reaches the camera sensor and for how long, which affects the image's exposure and depth of field.

A lower f-stop number (like f/1.8 versus a higher value, like f/11) allows more light to enter the camera. In general, use a lower f-stop number in a low-light situation or to capture an image with a shallow depth of field.

Applying the Pro settings

When you know how to coordinate the universal Pro settings (presented in the preceding section), you can produce some amazing photographs under difficult conditions. For example, the northern lights are tricky to photograph, mainly because you're in the dark and the lights move quickly. I used Pro mode to shoot photos (which I love!) of the *aurora borealis* in Alaska, as shown in **Figure 12-7**.

If you're interested, you can find the entire *aurora borealis* series on my Flickr account, at `flic.kr/s/aHsmAK9FPK`.

Photo courtesy of © Marsha Collier

FIGURE 12-7

When you capture a particularly good image, you can find the Pro settings you used when taking the photo, by referring back to the photo's details, or *metadata:* Go to Google Photos, select that fantastic shot, and swipe upward to read the metadata. For example, the screen shot in **Figure 12-8** shows f-stop, ISO, image size, and location for the photo shown in this section (refer to Figure 12-7).

REMEMBER

You can also find Pro settings metadata for photos from your smartphone's gallery. Tap the Gallery icon, select the photo, and tap the I icon (for information) or the three vertical dots (to access Details) for the photo.

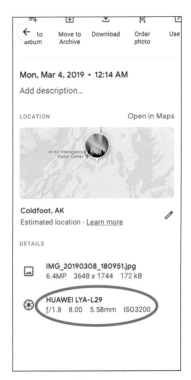

FIGURE 12-8

Edit Your Photos

Sometimes when you take a photo or video, even with the best of efforts, it's just not quite right. Today's smartphones all have useful editing capabilities built right into the phone. Each phone brand has its own spin on editing, but they all cover the basics you need.

Finding editing options

To edit a photo, tap the Gallery icon and follow these steps:

1. Tap on an image in the gallery or, if the gallery opens to *Album* mode, select Camera and then tap an image.

2. When the image opens, you may (depending on your phone brand) see icons for tools and options on the screen, as called out in **Figure 12-9**.

OnePlus 9 Pro 5G

FIGURE 12-9

These tools and options work this way:

*   **Google Lens:** Tap this tool's icon to identify items (such as faces) in your photos. Refer to the earlier section "Expand your reach with Google Lens" for more information on this tool.

- **Favorite (Heart ♡ icon):** Tap the hollow heart symbol; It turns into a solid heart (♥), and Google makes note of your favorite and puts it into a Favorites album.

- **Edit (Pencil icon):** Tap here to open the Gallery app's Photo Editor.

- **Delete (Trash Can icon):** Tap here to delete the photo from the Gallery app.

TIP

If you delete a photo accidentally, don't fret: Most Gallery apps have a folder called Recently Deleted (or Trash folder), where your device stores deleted photos for 30 days before their eventual destruction. Locate the deleted photo in that folder to restore the image to the gallery. You may have to access this by tapping three vertical dots in your Gallery.

- **Share icon:** Tap the Share icon to share a photo to any app, email, text message, social media post, phone, or other device — even a printer. The options seem limitless. Share is your friend, and you can find this universal icon — three dots with adjoining lines — all over any Android device.

3. If you want to edit the chosen photo (or video), tap the Pencil icon, and a new screen appears, showing the editing tools. The Gallery app's tools for editing are shown in **Figure 12-10**.

Applying the photo editor's tools

After you tap the Pencil icon, you find what may be a dizzying array of editing tools. To be honest (or, as the kids say nowadays, TBH), unless you know what you're doing, some editing tools may cause more harm than good. Practice makes perfect, and refining your editing talents takes a bit of work.

REMEMBER

As long as you haven't saved an image change, you can always go back to the original and try again. If you aren't happy with the adjustment from your edit, tap the X (instead of the check mark), and the photo reverts to the original (pre-edit) version.

— Editing tools

Photo courtesy of © Marsha Collier

FIGURE 12-10

Here's a description of the three basic tools that I believe you'll find valuable in the photo editor:

» **Crop:** Tapping the Crop icon enables you to adjust the viewable portion of the image. Tap to grab the dots at the corner to home in on what you want to be the focus of the picture.

» **Rotate:** We've all had that sideways photo that won't automatically jump to the expected rotation point. Tap the Rotate icon, and your image will spin 90 degrees each time you tap to its final position.

» **Mark:** Some folks enjoy drawing or adding text to a picture. Tap here to open a menu of options.

You'll likely notice some more advanced tools as well.

Playing with filters, colors, and more

Tapping the Filter option (icons for this feature vary among phones) gives you a slider menu with samples of varying color filters that you can apply to a photo, as shown in **Figure 12-11**. Tap one to try it out or try them all. You never know which filter may add that extra bit of magic to an image.

Photo courtesy of © Marsha Collier

FIGURE 12-11

The same goes for the tools you can adjust with a slider — for example, Brightness. Slider controls start in the middle as a point denoting the picture in its current version. By moving the control one way or the other, you can alter the look of the photo. Slider controls work for editing exposure, contrast, color saturation, shadows, warmth (using the Kelvin scale that I mention in the earlier section "Discovering the Pro settings"), and others.

Samsung photo editors do a great job. I used the default Magic Wand on the photo shown in **Figure 12-12** — and then the slider control showing the Kelvin scale by degrees.

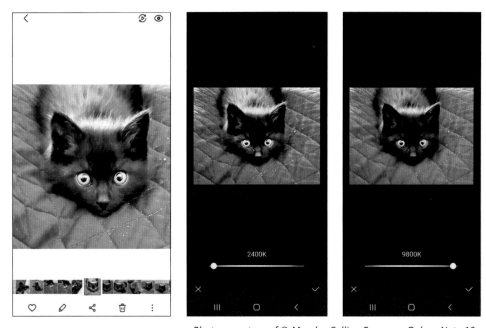

Photos courtesy of © Marsha Collier; Samsung Galaxy Note 10+

FIGURE 12-12

When I'm working on a masterpiece, rather than just fix photos on the fly, I wait until they sync with Google Photos in the cloud. Google Photos on your PC has a *robust* (yes, even more editing tools) editor that you can use on your phone or desktop computer (with your mouse). When you subscribe to Google One (see Chapter 1), you get even more bonus tools.

4

Exploring Android Apps

IN THIS PART . . .

Uncovering fun preinstalled apps

Discovering Quick Settings

Managing and editing with Google Photos

Learning to love Google Maps

Finding a missing phone

Establishing your apps' permissions

Enjoying entertainment on your phone

Chapter **13**

Preinstalled Tools You Want to Use

The folks at Google recognized that some built-in Android tools are so popular that they need shortcuts. Swipe downward on the Home screen, and on the resulting window shade you can see the shortcut icons to access Quick Settings tools. Swipe downward again, and you see even more icons and the tools' names below them.

These shortcuts to the phone's settings enable you to make the changes you need on the fly with a swipe and a tap. You can also find all these tools and their settings by searching the main settings, accessed by swiping downward on the Home screen and tapping the cog on the window shade.

I talk a bit about the Quick Settings in Chapter 7, but in this chapter, I present *the* more useful tools and help you, I hope, feel comfortable with them. The specific tools that appear in the shortcuts depend on your brand of phone and on the version of the Android operating system (OS) you have. A short list of these preinstalled tools, or features, appears in this chapter, in no particular order.

Take a Shortcut to Features with Android Quick Settings

Using the Quick Settings for the handy preinstalled features doesn't involve a lot of complexity, so this section offers some basic instructions for how to work with the features and their settings in the shortcuts area. It's handy to have these settings readily available, and I like working with settings this way.

REMEMBER

Which Quick Access settings appear in the shortcuts area on your phone is based on the phone's manufacturer and its Android OS version. *Note:* Your phone may add some manufacturers' settings here as well.

Customizing the Quick Settings

Follow these steps to start working with the Quick Settings:

1. Swipe downward on the Home screen from the icons at the top (time display, battery percentage, and Wi-Fi connection, for example). Swipe downward a second time to see a full screen of Quick Settings icons.

2. Below the bottom row of icons, you may see some dots, as shown in **Figure 13-1**. Swipe right to reveal more screens of settings icons. (You find as many screens of icons to review as there are dots at the bottom.)

3. To activate or deactivate a feature, tap its icon to turn it on or off.

TIP

Long pressing on many of the icons can open up the app's full settings menu from the System Settings. There you find more options for the feature; versus just the on/off toggle.

4. To move, add, or remove a setting in the shortcuts area, tap the Edit icon (the small pencil) at the bottom of the first screen (refer to Figure 13-1). On other phones, you may tap a plus (+) sign in a circle.

5. In Edit mode (see **Figure 13-2**), select the setting you want to change by holding your finger on it.

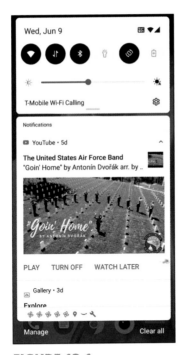

FIGURE 13-1

FIGURE 13-2

You can then perform the following actions:

- **Move a setting:** Hold your finger on the icon and move it to the position where you want it to appear.

- **Add a setting:** Below the active settings you see a small group that is grayed-out, as shown toward the bottom of Figure 13-2. To activate one of those settings, tap and drag it into the active settings area.

- **Remove a setting:** Hold and drag the icon you want to remove down into a blank spot in the grayed area. Then lift your finger to drop the icon into the inactive zone, maybe labeled Drag Here to Remove.

TIP

On a Samsung phone, you can tap the three dots in the upper right corner on the main Quick Settings screen to see a drop-down menu and choose the option to enter Edit mode (see **Figure 13-3**).

Samsung Galaxy S10+

FIGURE 13-3

Meeting popular Quick Settings

Table 13-1 presents the most common Android features in Quick Settings.

TABLE 13-1   Main Android Quick Settings

Icon	Name	What It Does
●	Wi-Fi	Turns Wi-Fi on or off. When Wi-Fi is on, your phone connects to a network that you store with a password, tap the icon to turn off Wi-Fi. If you want to change networks, long press on the icon to see a drop-down list of available connections. (Leaving your Wi-Fi turned on when you are not in a Wi-Fi zone can cause battery drain as the phone searches for signals.)
⇅	Mobile Data	Turns on or off the data use from your mobile carrier. At home, you're probably already paying for Wi-Fi (and it's as fast as you need), so why not use it by turning off mobile data? Google Voice and some mobile carriers also offer VoIP (Voice over Internet Protocol) at no extra charge, which is the technical term for making voice calls over Wi-Fi. If you have a plan that charges extra (sometimes, crazy extra) for international roaming, you can turn off data here. ***Note:*** Currently, T-Mobile is the only carrier (in late 2021) that includes free international texting and data in its plans.
✳	Bluetooth	Toggles the Bluetooth connection on or off. If you use devices (such as headphones, earbuds. speakers, smartwatches, or fitness bands) that connect to your phone wirelessly, tap the Bluetooth icon to toggle this feature on and off. Long press on the icon to pair (connect) a new device.
✈	Airplane Mode	Toggles the connection to wireless transmission of phone calls. Activating Airplane mode (toggle on) cuts the wireless connection per Federal Aviation Administration guidelines. You can still use the plane's Wi-Fi to watch movies or listen to music.

(continued)

TABLE 13-1 *(continued)*

Icon	Name	What It Does
	Hotspot	Turns your phone into a hotspot to share your data signal. If your cellular plan includes hotspots, your phone can become a hotspot to share your unlimited cellular data signal with friends. They sign in to your phone just like it's a Wi-Fi network.
	Data Saver or Mobile Data	Turns on or off app usage of data. On a limited data plan, you may not want your phone using data without permission. Turn on this setting to prevent apps (like Gmail and others) from randomly updating.
	Location	When toggled on, this setting reports your phone's location to apps you're using (such as Google Maps). It may be critical for emergencies because the FCC is requiring more 911 responders to find your location. For safety's sake, I keep this one turned on. **Note:** Unless you're using a GPS app (like Google Maps), this won't drain your battery.
	Auto Rotate	When toggled on, switches between Portrait (vertical) and Landscape (horizontal) modes depending on how you hold the phone. If you aren't able to rotate the Home screen to horizontal orientation even with this enabled, try long pressing on a vacant space on the Home screen to get to Home settings and check whether your phone has the option. **Note:** Don't touch the screen when you want it to rotate.
	Battery Saver	Turns on or off a power saving mode if your phone's battery level is creeping lower and lower and you can't get to a charger. If a draining battery happens to you a lot, I highly recommend that you carry a small charger. (I explain more in Chapter 16.)
	Flashlight	When toggled on, it turns your phone's camera flash into a decent flashlight. So handy!

Icon	Name	What It Does
◉	Screen Recorder	Toggles on or off the capability to archive anything that appears on your screen. This feature is different from a static screenshot; it enables you to record a chat conversation, for example. The recording becomes a small movie stored on your phone.
☾	Dark Mode	Toggles the phone screen between modes. The On setting turns your phone's screen to a black background with white print on many apps. This setting saves on battery use, but can be bad for your eyes, which need more light for reading.
⤢	Nearby Share	Enables (or disables) a link for sharing with a nearby Android user's device via Bluetooth or Wi-Fi. Like Apple's AirDrop, use Nearby Share whenever you want to share files, photos, or a story link with a nearby device. Nearby Share is an alternative to sending media via text message or email.

In the phone's main settings for Nearby Share (you can get there by long pressing on the icon in Quick Settings), you can set Nearby Share to be available to all contacts or some contacts, or to be invisible. When you attempt to share by using the Nearby Share icon, your phone asks whether you want to turn on the service.

TIP

As a matter of security, the better option for Nearby Share is to approve sharing each time you use it. When you attempt to share with another Android phone, your screen shows the handshake (nearby connection) occur (see **Figure 13-4**), and you can choose to either Accept or Decline the share.

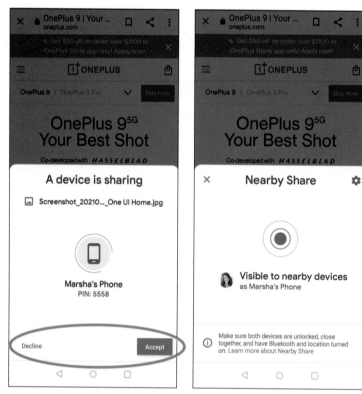

FIGURE 13-4

Quick Settings That Offer Valuable Options

Many Quick Settings are simple on–off toggles, but some are more flexible and give you options. I use all the Quick Settings listed in this section, and I show you how to adjust them to your lifestyle and preference. Press and hold your finger on the icon to get to the settings' options screen.

Opting for Dark mode

I mention Dark mode in Chapter 6 in relation to the appearance of the keyboard. Whether you want to read white text on a black background or read black text on a white background (see both options in **Figure 13-5**) is a personal choice. As far as your eyes go, I'm told that black text is easier to read. But at night, I prefer the Vision Comfort setting (see the later section "Maximizing eye comfort") to lull myself to sleep while reading from my phone (or tablet).

Maximizing eye comfort

In my experience, Vision (or eye) Comfort is one of *the* greatest smartphone innovations. Invoke Vision Comfort, and magically your eyes feel better. When turned on, Vision Comfort filters out blue light, and the resulting screen takes on a somewhat yellowish tint. This mode is the best one for long reading sessions on your phone.

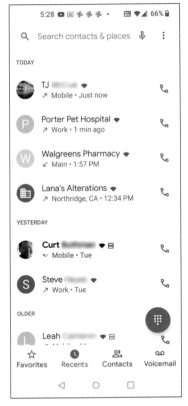
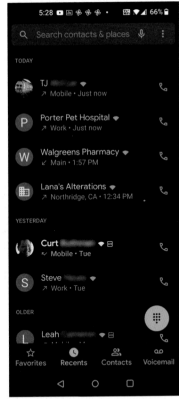

Default mode   *Dark mode*

FIGURE 13-5

After you long press and hold on this setting's icon, you find sliders (as shown in **Figure 13-6**) that adjust the screen for your best comfort levels. Place your finger on the dot and move it back-and-forth along each slider to see how the adjustment affects the look of your phone's screen. Don't rush the process. Take your time to find adjustments that give you a functional, readable screen — without burning your eyes. You can always alter this setting.

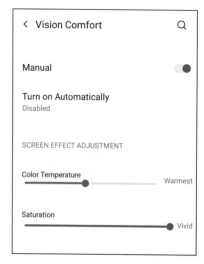

OnePlus 9 Pro

FIGURE 13-6

TIP

If you think that more options may be (or should be) available for a particular tool's settings, don't be shy about long pressing any other toggle or link you see in the settings area. You may be surprised at how deeply you can refine these preferences.

Avoiding interruptions

Okay, maybe Do Not Disturb is the best setting ever! Tap the icon to turn on this setting, and your phone won't make a peep (until you tap again to turn it off). It's perfect for those times when you — temporarily — don't want to be disturbed. Tap and hold the icon to bring up more controls.

As **Figure 13-7** shows, you can set a schedule for which hours (maybe overnight or during work) that you want to designate as your phone's off hours. For example, set the schedule for the period when you don't want to be disturbed while sleeping.

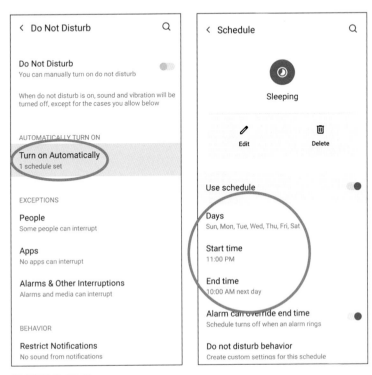

FIGURE 13-7

To schedule Do Not Disturb automatically, follow these steps:

1. Long press the Do Not Disturb icon.

2. On the resulting screen, tap Turn On Automatically. Sometimes this option reads *Turn On as Scheduled,* and you can customize when you want the automatic on to happen. Look for a + sign to add a new schedule.

3. Tap the settings cog next to the word *Event.* Then choose an event from your calendar, and the phone will remain quiet during that time.

4. For peace during the night, tap *Sleeping* and then tap the little cog to the right of that word. Refer to Figure 13-7 for the options you can set (Days or Start Time, for example) and the Edit icon you tap to set them.

   By tapping the various options, you can set a schedule with a sleep time to shut off phone interruptions and a wake-up time to allow your phone to come back to life. You can also indicate on which days the schedule is active.

Fortunately, you can set exceptions to a Do Not Disturb schedule that allow certain people from your contacts list to get through at *any* time. Another option enables you to permit a call if the same person calls a second time within a 15-minute period. These exceptions are shown in **Figure 13-8**.

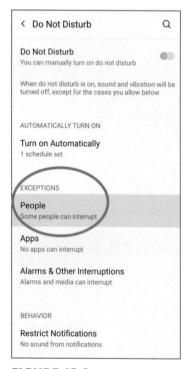
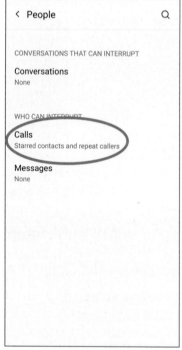

FIGURE 13-8

Samsung phones' Do Not Disturb Settings screens look a bit different but have the same basic settings.

REMEMBER

It's pretty funny when the wake time rolls around for me. I'm not an early riser, so when my phone wakes up, I don't need an alarm, I hear the "ding-ding-ding" indicating texts have come in while the phone was in Do Not Disturb mode. (I'm not *that* popular, but my supermarket, Amazon's Treasure Truck, and others text my phone regularly with their deals of the day).

Sharing with "close" friends

While writing this book, I used Nearby Share (just like Apple's Air-Drop) almost every day. See what sharing looks like in **Figure 13-9**. You can use Nearby Share to send photos, news links, videos (almost anything on your phone) with someone else who also has an Android phone in the same room. It works seamlessly.

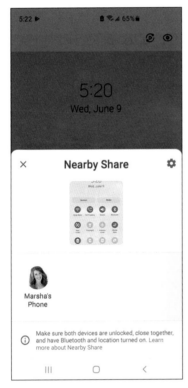
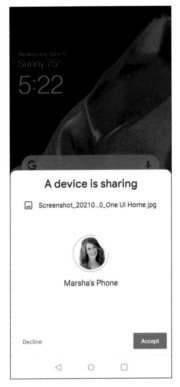

FIGURE 13-9

Phone-Resident Android Apps

Without your needing to download any apps, your phone has many valuable apps of its own to enhance your smartphone experience. Your phone (as you may now know) has so many functions! This section notes a few resident apps that are worth using.

Recording your voice

The ability to record your own voice (or others) is darn useful in many ways. You can record a lecture you're giving or a conference address. If you're a musician, you can record ideas that come into your head — the applications are limitless.

Before you go to the Google Play Store to look for a recorder, a perfectly simple-to-use app is installed on Android phones. It's called Recorder, and on Samsung phones, Voice Recorder. You can find it in the Apps drawer. *Note:* When your apps are sorted alphabetically, it's easier to find what you're looking for.

Call recording *is* a built-in feature of Android phones, but the capability is restricted by carriers, manufacturers, and by geographic zones. You don't need many instructions to use the app. Just open the recording app and tap the Microphone icon to start the recording.

Whenever you're ready to end the session, tap the small square that appears during recordings. After the recording is finished, you can also rename the audio file in the app so that you can identify the record of what you've recorded.

TIP

I recommend using your phone's recorder when you're on a customer service or tech support call — I always do. I place the call on a land line (or a friend's phone) using the speaker mode and put my phone next to it to record. When the company's recording announces that the call may be recorded for training purposes, I jump in and say, "I will also be recording the call so that I'll remember everything we speak about."

REMEMBER

When you record a phone call, it's important to let the other party know that you will be recording in case you ever need to use the recording in small claims court.

WARNING

I know that there are laws regulating the recording of conversations. California, where I live, is a two-party state. That means that both parties must give consent to the call being recorded. Indiana is a one-party state; meaning that you can surreptitiously record a call without the other party's knowledge. Search Google for your state's laws regarding recording phone calls or conversations.

In the settings area (as always, look for the small Cog icon or the three vertical dots to get there), the Samsung Voice Recorder has an added feature. You can convert up to 10 minutes of speech to text (see **Figure 13-10**). You get all this for free on your phone.

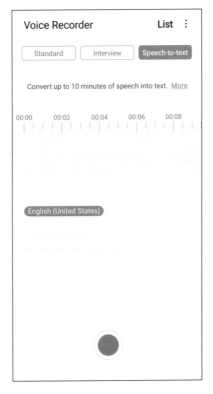

FIGURE 13-10

In Chapter 14, I tell you about Google Keep, a note-taking app that allows you to record short notes for yourself (and retrieve them on any of your Android devices, on iOS, or on your PC's Chrome browser).

Taking a screen shot

Sometimes you just want to take a *screen shot* (it's basically a snap-shot) of the content on the phone's screen. Perhaps you want to save a funny meme, show proof that you paid a bill, or even capture a Facebook post. Whatever you want to keep, you can save it in a screenshot.

Here's the rub: Each manufacturer may have extra methods of taking screen shots. Some have a Screen Shot icon in the Quick Settings and others — well, here are a few variations:

» **Native Android method:** When you press both the power button and the Volume Down button at the same time, the phone briefly flashes, and you may hear a click. These actions let you know that you've successfully taken a screen shot. Also, if Google Assistant is installed on your phone (if not, I would download it *right away*), just say, "Hey, Google — take a screen shot," and it will do as you say. (Find more about Google Assistant in Chapter 7.)

» **OnePlus:** You can use the native Android screen shot method, but OnePlus adds an easy-to-use feature: Tap Settings ➪ Buttons & Gestures ➪ Quick Gestures ➪ Three-Finger Screenshot and toggle it on. Then drag three fingers from the top of the screen to the bottom. Your phone captures the screen shot and shows you a preview version at the bottom of the screen. Tap the preview version, and you can edit it by cropping, drawing, altering colors, and more. If you miss the preview (it doesn't appear for long), you'll find the screenshot in your phone's Gallery. Just tap it to edit.

» **Samsung:** Most Samsung phones respond to the standard Android method (pressing both the power button and the Volume Down button at the same time). But some Samsung phones offer another way: My favorite is to tap Main Settings ⇨ Advanced Features ⇨ Palm Swipe to Capture.

When you turn on this feature, you can swipe the edge of your hand across the screen to take a screen shot. A small version of the screen shot appears on a toolbar at the bottom of the screen. Tapping there enables you to edit the image.

» **Huawei:** Huawei phones have the same default screen shot method as most Android phones — but they add a new gesture to the mix. Search Take Screenshot from the Main Settings icon or go to Settings ⇨ Accessibility Features ⇨ Shortcuts & Gestures ⇨ Take Screenshot ⇨ Knuckle Screenshot. You see a page where you can toggle Knuckle Screenshots to On (see **Figure 13-11**). Afterward, all you have to do is knock twice on the screen and you see a screen shot.

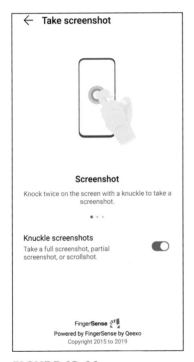

FIGURE 13-11

REMEMBER If you're looking for more built-in screen shot options for your phone, ask your carrier or go to your manufacturer's Help page. You can find a table with the manufacturers' customer service links and phone numbers in Chapter 1.

Staying on task with Google Calendar

I know. It's a calendar — big deal. But this calendar can help you keep track of your appointments as well as keep up with others in the family. After you add an event to the calendar, you can invite others to it and receive reminders ahead of time.

A Google Calendar event is shareable with your family (and friends), if you want to do so. When you add someone to an event, Calendar sends them an invitation. When they supply a Yes response (as shown in **Figure 13-12**), the calendar entry appears on their calendar as well.

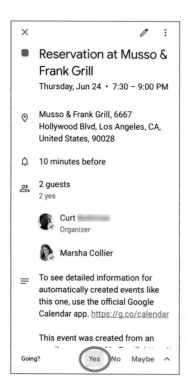

FIGURE 13-12

Google Calendar integrates with other apps, and my favorite integration (living in Los Angeles, where traffic is a life issue) is with Google Maps. When you enter a destination's address into the calendar, the calendar checks with Google Maps and suggests the best time to leave for an event. It's reassuring to know that you won't miss your dinner reservation across town because of traffic. I tell you more about Google Maps in Chapter 14.

Depending on your brand of phone, you may have the phone manufacturer's own brand of calendar app pre-installed. You can identify your app by clicking a Cog icon or three vertical dots within the app; you will know it's the Google Calendar app if it has your Google Account avatar in the upper right. If you prefer to use Google's app, you can always download it from the Play Store and use it instead as the default app.

Frequenting the Google Play Store

The Play Store is the official place for digital distribution of apps for your Android phone. It has over 3 million apps that have been carefully vetted by Google engineers. The apps should not cause any security threats, which is not to say that it's a perfect playground — but it's close. In Chapter 15, I cover several must-have Google apps from the Play Store.

You may find that your phone brand has its own app store and probably suggests that you download apps from there. I'm sure this suggestion is perfectly safe, but in the world of security, you're just opening up a new door to your data.

Make your decisions wisely — whenever you have to click to *agree* to go to a new place on the Internet or on your phone, you're agreeing to share your data with that party. If you don't agree, you won't be able to go there. You have already given permission to the Google world by purchasing an Android phone.

IN THIS CHAPTER

» **Retrieving Google apps**

» **Having fun with Google Photos**

» **Finding your way with Google Maps**

» **Locating a misplaced phone**

» **Keeping a digital notepad**

Chapter **14**

Google Mobile Services Apps for Android

The Google apps I present in this chapter may or may not appear on your phone, although most of them should. All part of Google Mobile Services (GMS), they're made for the Android OS (operating system). If a covered app did not come installed on your phone — and you like what it offers — you can download it from the Google Play Store.

As I mention in Chapter 5, phone manufacturers and broadband carriers may add their own apps to your phone. Google can approve these apps, but they aren't official Google (GMS) apps.

I believe that using GMS apps helps you keep your data reasonably safe and private. After all, iPhone users also have the App Store (the Apple version of the Play Store), but a good deal of the apps on their phones are native to the Apple ecosystem.

I just looked at a list of GMS apps — it's comprehensive! The apps I present in this chapter can help make your smartphone an all-in-one mobile communications center and office.

Find Popular Google Apps

You may be surprised at the number of GMS apps that are available, and I can't possibly list them all in this book. When you visit the Play Store, just search for the word *Google*. Most of the app names are prepended by the word *Google*, so by searching for it (as shown in **Figure 14-1**), you find a world of apps. Depending on the type of phone you have, you can see as many as 35 apps from Google.

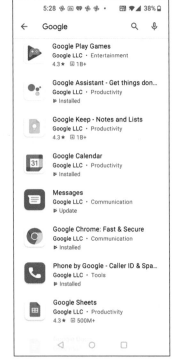
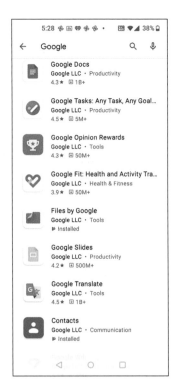

FIGURE 14-1

Notice that some apps are duplicates of ones you have on your phone (they're updated via the Play Store) and that others have an app icon that's just a little (or a lot) different. If you want to know whether an app on your phone is an official Google app, you can check by clicking around in the app's settings.

For example, you may see a generic Email app on your phone. That isn't Gmail (Google's official mail service, described in depth in Chapter 11), although you *can* use it to send and receive Gmail. Remember the following statements when you choose which email app to use:

» The email app that's preinstalled on many phones allows you to combine email addresses from Hotmail, Yahoo! Mail, and more. Samsung, for example, adds its own email app that enables you to add all your email accounts under this single app.

» The all-in-one approach is convenient, but keep in mind that you can find separate Android apps for Microsoft services and Yahoo! Mail, if you want to maintain separate email accounts.

» If your email address is from an Internet service provider (ISP), you have to use the generic email app unless you want to check your email via your phone's web browser.

In Chapter 1, I talk about the benefits of setting up a Google account. After you do so, your information syncs between all the apps and makes Android phone apps integrate smoothly and play well with other devices, like your PC, with a Chrome browser and home Internet devices.

Google Photos

Google Photos launched in 2015, and I jumped right on it. I downloaded the app and synced my phone to upload photos to Google. According to the Google blog, as of November 2020, users of Google Photos have stored over *4 trillion* photos on the platform. And every week, users upload 28 billion *new* photos and videos. Although the numbers are impressive, what they tell me is that I'm not the only one who's happy with the service.

Your phone manufacturer may also offer a cloud service for storing your photos, but the free storage space is probably limited. Remember that your Google account gives you 15 gigabytes (GB) of free storage to start out. If picture and location privacy is an issue to you, remember that you have an Android phone — Google is already up in your cheesecake.

ARCHIVING AND BACKING UP PHOTOS

As the Google Photos app takes on more and more features, I am even more sold on the platform. For example, I fired up my older computers and synced many of my older personal digital photos with Google. I had decided to make the platform my family photo archive. Maybe one day I'll digitize all the paper photos I have stored from decades past and upload those, too.

In addition to having an online archive, it only makes sense to keep a backup of your photos on an external hard drive at home. You can purchase a portable, 1TB (a *terabyte* equals a million megabytes) hard drive online for less than $50 to back up all the treasured photos from your devices and memory cards.

Remember: You can *never* have enough backups.

WARNING

Be sure to set defaults for your preferred apps as soon as you find other similar apps on the phone. Sometimes, especially on a recently acquired device, the manufacturer's clone photos app decides to upload your images (or files) without explicit permission. I've found my photos on random cloud drives, either duplicates of the images I have in Google Photos or originals. (I am trying to figure out how this happens and fear I'll waste a ton of time doing so.) Moral of this story? Establish and use preferred apps pronto.

Storing and retrieving

Google Photos is completely separate from your phone's photo gallery. (I talk about the photo gallery in Chapter 13.) When you install Google Photos, you need to tell the app to sync to Gallery and its subfolders as they appear. Your phone may set up separate folders for downloads from text messages, social media posts, screenshots, and more.

Maybe you already have a Google account and have been storing photos and videos in Google Photos for years — but you have no local backup. (See the earlier sidebar, "Archiving and backing up photos," for a look at my experience.) What you upload is *your data*, and Google makes sure that you have access to all of it. You can use Google Take-out to download a copy of the data from Google Photos or any other Google products you use.

TECHNICAL STUFF

Go to `takeout.google.com` and select Google Photos from the list of items you can retrieve. Follow the steps on the page, and you incrementally receive multiple .zip files (or download links) with your photos and videos.

Syncing and deleting

A word about syncing: When you set up Google Photos to sync online, everything you mark for syncing uploads from your phone's Gallery app. For example, all screen shots of my phone's screen that I've taken for this book upload to my Google Photos. If I delete them in my smartphone's gallery, they're still stored in Google's cloud.

To delete photos and videos you no longer want to keep, go into the Google Photos app (or go to `photos.google.com` on your PC) and delete them from there. You can delete from your phone, but it's so much easier on a big screen. You can delete multiple photos by selecting each one you want to delete and then tapping the Trash can icon. You select the photos by long-pressing with your finger on a touch screen or by clicking your mouse. A pop-up warning lets you know that you're deleting these files from your device's gallery at the same time.

REMEMBER

Cloud Trash cans empty every 30 days. So, if you make a mistake and delete something unintentionally, you can go to Trash and restore it.

Sharing

Where Google Photos absolutely *shines* is in sharing your photos. (iPhone users can use AirDrop — it's the same as the Nearby Share feature in Android described in Chapter 13 — or can email and text photos.) In Google Photos, you can share a single photo or even an entire album by sending a link that's generated from your Google Photos account. You select a photo by *long-pressing* it — press it and let your finger linger on it for an extra second — until it sports a tiny white-on-blue check mark in the corner. Then click the Share icon. The photo or album you select to share is available to a third party only from that link.

Searching

The *best* feature of Google Photos is its Search function. After you upload photos to Google, the software identifies characteristics of the photos — what or who is pictured and where the photos were taken.

Figure 14-2 shows a Google Photos search on my phone for the location *Paris*. **Note:** I have my private timeline engaged so that each spot on the map of Paris links to a specific photo taken on that spot anytime I am in Paris. This automatic linking happens because I keep my phone's Location History setting toggled on. You can enable this timeline view by tapping the three dots in the map's upper right corner.

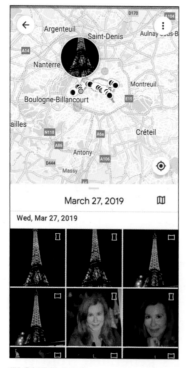
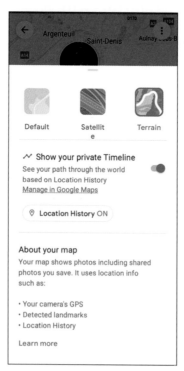

FIGURE 14-2

As shown in **Figure 14-3**, I have albums in my Google Photos app. I didn't make these albums (although I could); Google's artificial intelligence (AI) sets up albums over time, by grouping photos based on their characteristics. A little help is needed from you, too.

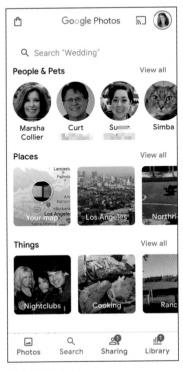

FIGURE 14-3

When you see a photo of a person or a place or an activity that you want to identify for your albums, tap the photo to enlarge it and then swipe upward. **Figure 14-4** shows a photo of our family's kitten, Sam. If I swipe upward on the photo, an entirely new screen with options appears. You can add information, such as a description of the photo, from this screen. (If Google Photos already knew that the kitten's name was Sam, the text *Sam* would appear above the thumbnail version of the photo on the bottom left.)

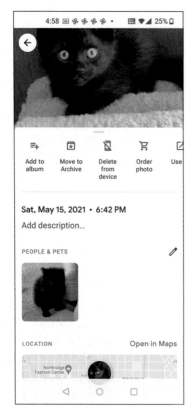
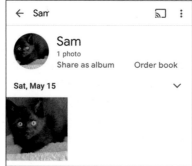

FIGURE 14-4

To add information that identifies a photo for searching after you swipe upward on the photo, follow these steps:

1. Tap the words *Add description* and type an easily searchable keyword description of the photo as prompted.

2. Tap the Pencil icon and then tap the face of the person (or pet) you want to add a name for. Tap the plus sign (+) and confirm that you want to add this face for recognition.

3. Tap the thumbnail of the face and, on the resulting screen, you're prompted to add a name.

To add Sam to my photos, I tap the Pencil icon and select his face. Then I go back to the thumbnail and tap it. Figure 14-4 shows (on the right) that I added Sam's name to his thumbnail. Now Google finds all my photos of Sam and makes them searchable by his name.

4. If you didn't have Location History enabled when you took the picture. The location tag will be blank. To add a location, tap the location and make the changes.

I have taken tons of pictures of my cats, and it's fun to look back on their development and growth. Searching for people is fun, too. I can find every photo of my husband or daughter by merely typing their name into the app. Google also finds any untagged photos and matches faces to identify them.

Google Maps

You *must* have Google Maps! Back in the olden days (just a couple of decades ago), everyone used to shake out sprawling fold-up maps and wrangle thick paper atlases in their lap in order to find their way around town. (This experience was *nerve-wracking* in a tiny vehicle.) Now you just say or type an address, or click an address from a calendar event, and a detailed map shows up — pronto! — to direct you.

Finding your way to an appointment

When you set up a calendar appointment with a location, you see the address next to the Pin icon. Open your calendar and tap the appointment to see these details. **Figure 14-5** shows how the address link and Pin icon appear.

Tap the address link to see the location on a map. As shown in **Figure 14-6**, the details at the bottom of the screen tell you how long it takes to drive to the location. (And you can sometimes preview the parking situation.) Places you have visited previously often show up on the map — which is nice for visual reference.

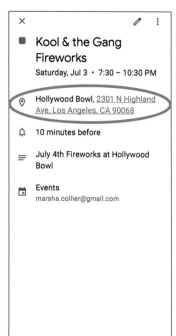

FIGURE 14-5

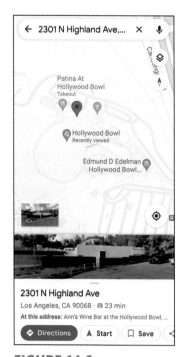

FIGURE 14-6

When you tap the Directions button, the map changes to a point-to-point map. In the point-to-point version, you can tap to select the transit time (and route) for driving, walking (six hours — wow!), or public transport and rideshare times. On some routes, (indicated by a small green leaf) Google suggests a more fuel-efficient route as well. You may also see a slightly slower alternative route to reach your destination.

To follow directions from your smartphone, place the phone in a phone holder (or cupholder) and tap the Start button. The position indicator (a blue and white circle) follows you along the way, and Google Maps speaks out the turn-by-turn directions.

Employing the Directions screen options

Before you're en route, you can tap the three dots in the upper right corner of the Directions screen to reveal several options. The options I use most from the Directions screen are described in this list:

» **Add Stop:** This option opens a new text block, where you can add an address. Press the arrows that appear on the side to change the order of the stops.

» **Set Depart or Arrive Time:** Set the time of your appointment on the clock (see **Figure 14-7**). Select Depart At or Arrive By to coincide with the time you set, and then swipe to select the date of the event. (I personally put the time I want to arrive in this spot.) When you finish adding arrival-time data, Google's AI finds the right time for you to leave. I've found this feature to be more accurate when I wait until the morning of the event or appointment to use it.

» **Share Your Location:** You can select a contact with whom to share your location for a prescribed period (or forever). I know this sounds weird, but I use it for tracking when I travel. For a trip to the airport in Shanghai (where I couldn't read any of the road signs), it was nice to know that the driver was taking me directly to the airport. And my husband knew exactly where I was.

FIGURE 14-7

Find My Device

Using the Google Find My Device feature is short and sweet — and you don't even have to download an app (though you can, if you want). Do you ever misplace your phone or tablet or another Android device? I admit that I do — hey, it happens.

As long as you have access to another Android device linked to your Google account and a Wi-Fi connection *and* you have your lost device synced with Wi-Fi and your Google account, you can find the errant one. Unlock the device you're using (tablet, computer, or another phone) to locate the missing one, and then follow these steps:

1. Type this address in the Google search bar: `www.google.com/android/find`. You should automatically see a map showing the location of your missing phone, as shown in **Figure 14-8**.

— The missing device's location

FIGURE 14-8

2. If you have more than one synced device, you can tap one of the miniature phone versions at the top of the screen to choose the lost one.

3. With the missing device chosen, you can tap to play a sound from the device, lock it, or — if all else fails — erase the device entirely (only if you believe that it has fallen into questionable hands).

TECHNICAL
STUFF

Being able to find a lost device is an excellent reason to hold on to an older (retired) phone and connect it to your home Wi-Fi. You'll not only have another device kept safe at home but also be able to use it for other tasks, like finding your phone when it wanders off (and watching YouTube videos, of course). You can also use it to make Google Voice calls using Wi-Fi.

If you want a Find My Device app (I use my browser rather than the app, to save space on my phone), it's in the Play Store. Beware of imitations — copycat apps abound! Download *only* the official Google Find My Device app from Google LLC. It has over 100 million downloads.

Take Note(s) with Google Keep

Google Keep is a place to keep lists and notes that sync to all your devices. This app is simple and easy to use. Download it from the Play Store and spark it up! Tap the colorful plus sign (+), as shown in **Figure 14-9**, and type your notes.

I use Google Keep to jot down recipes, ongoing shopping lists, notes for my doctor, and the titles of books and movies I want to buy or check out of the library, for example. I find this app to be invaluable, and I can edit or read from any device connected to my Google account (including my laptop at `keep.google.com`).

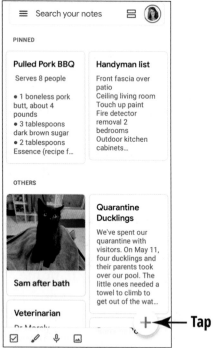

FIGURE 14-9

IN THIS CHAPTER

» Setting app permissions

» Listening to radio, podcasts, and music

» Watching TV and movies

» Gaming by yourself or with others

» Reading news or books

» Engaging with social media

» Examining Android accessibility apps

Chapter **15**

Apps You Might Like in Google Play Store

hroughout this book, I mention apps made by Google that can enhance your smartphone experience. But more than just Google apps are popular — and for good reason. They work well and possibly replicate a nondigital tool (such as an authentic paper newspaper) that you may have used for decades. Apps make these familiar tools super portable, which can in turn make life easier (or at least lighter).

Whenever you download and use any kind of app on your phone, you need to consider privacy for the information associated with the app. Luckily, in Android 11 and 12, you have much more control over which apps can access what information. Early in this chapter, you find out how to carefully establish your apps' privacy permissions.

For input to this chapter, I asked many friends about which apps they use regularly. Here's a round-up of some must-have items from the Google Play Store.

Establish App Privacy Permissions

When you install a brand-new app from the Play Store, you have options for accepting or denying that app permission to access information on your phone. For example, during installation, an app may ask for permission to use your camera, contacts, location service, or microphone.

It's a myth that you have to agree to all these permissions. I suggest that you agree to ones you feel comfortable with and see whether the app accepts your permissions choices — and functions well using them.

TIP

A recent Play Store services update began rolling out for devices running Android 6 Marshmallow and higher (which includes billions of phones). In a massive win for privacy, the update gives your device permission to *auto-reset* permissions for apps that haven't been opened in a while. When reset, these apps no longer have open access to your sensitive personal data, including tracking your location, camera, contacts, files, microphone, and phone. You must then re-grant permissions if and when you reopen the app. (Why not just delete it?)

As of this writing, Android 11 still represents only about 28 percent (and Android 10, 32 percent) of the current Android handsets in use worldwide. If you're using Android 11 or higher, you have at least these options to select from:

» **Allowed All the Time:** Selecting this option means that the app has access to the associated information or service all the time, whether the app is open or closed.

» **Allowed Only While in Use:** The app only has access to, say, the camera when the app is open and remains accessible until you close the app by flicking it off the screen. (This option is my preference.)

» **Denied:** Use this option when you never want to give permission for the app to access the information or service.

If you're using an older Android phone or you have restored apps from an older phone to a new one, you might want to take a look to see whether app permissions need to be updated. Unfortunately, there's no way to universally turn off permissions for previously installed apps — you need to visit each app individually.

1. To access the permission manager, swipe down on the Home screen and open the main settings by tapping the Settings cog.

2. Depending on your device, you usually tap one of these options:

- On Samsung: Privacy ⇨ Permission Manager

- On One Plus: Apps and Notifications ⇨ App Permissions

3. Tap Permission Manager (if needed) to see all the permissions granted to apps on our phone. You'll have to go through each one to make edits.

The resulting screens should look something like the screens shown in **Figure 15-1.** And now you can access all permissions granted to any app installed on your phone.

4. Tap on a permission (such as Calendar), one at a time, to see the apps being granted that specific permission. If needed, you may be able to modify the permissions to another option. Some apps may protest, but I've found many still work fine after denying permissions:

- **Allow All the Time:** So, this app is always watching?

- **Allow Only While Using the App:** This option makes sense and is my choice.

- **Ask Every Time:** You must respond each time an app needs access to something.

- **Deny:** This one may disable the use of the app. But it may not. Give it a shot? After all, why would Amazon Shopping *need* to know my Physical Activity?

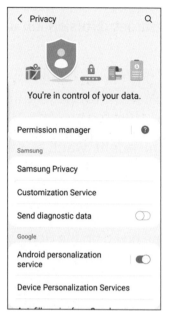
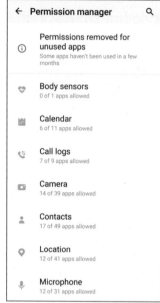
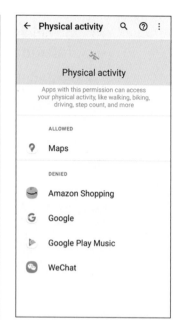

FIGURE 15-1

Connect to Radio (Yes, Radio), Podcasts, and Music

Did you enjoy listening to radio in the old days? You can still listen to it. All you need is a cellular connection (preferably, on an unlimited plan) or a Wi-Fi connection. These days, the Internet is full of entertainment, and your smartphone puts it in the palm of your hand.

You can listen through your phone's speakers or external speakers or a wireless Bluetooth connection to earbuds or a headset.

Configuring your speakers' volume

When you want to listen to music, a podcast, or radio through your phone's speakers, that connection happens automatically — when you turn on the app. The sound may be too loud or not loud enough, though. To control the speaker volume on your phone, look for the two buttons next to each other on the side of the phone. (You might find a rocker-style button instead; Chapter 5 explains about the volume buttons.)

Press the top button to increase the volume, and press the bottom one to decrease it. After you press either button, the volume control shows up on the screen, looking similar to the control shown in **Figure 15-2.** You can tap or slide the control to adjust the volume more precisely.

FIGURE 15-2

TIP

When your phone is connected to earbuds or external speakers, you can still use the phone's on-screen controls to fine-tune the volume.

Making a wired connection

If you have an older phone, you might be lucky enough to have a fitting for using an old-fashioned wired connection. I say *lucky* because a wired connection (for example, wired earbuds) gives you one less gadget to keep charged. Look for a small, round hole (the headphone jack) at the bottom of your device.

TECHNICAL STUFF

Newer phones may use a wired connection through the USB-C power connector with a USB-to-3.5mm audio cable adapter. If you want a wired connection, access your phone's documentation online at GSMArena (www.gsmarena.com). On the website, look up your phone by brand and model number. Go to the Sound category and look for a 3.5mm jack connection.

Some people keep an older phone around, rather than trade it in, and use it like a clock radio — with a tiny, wired speaker or just the phone's (possibly stereo) speakers. Many major brands still sell wired portable speakers for just this purpose. You can easily find them online for as low as $10 along with in-ear wired headphones and earbuds.

TIP

You can repurpose an old phone in a multitude of ways — for example, by making it your bedside clock-radio. As another example, turn an old phone into a security camera or just use the camera with Wi-Fi to transfer photos.

Using a Bluetooth connection

The *Bluetooth* wireless standard can send data over a short distance (in practical use, approximately thirty feet). It has come a long way since you may have tried it in the past, when connecting was iffy. Connections are quicker now, and unlike the situation a few years ago, you can connect multiple devices to your phone's Bluetooth connection.

TECHNICAL STUFF

Bluetooth is a safe radio frequency (RF) that operates in the 2.4GHz range. It is said to be a thousand times weaker than a cellular connection (hence, the short distance it reaches). It is also secure once the connection is made — and it's private.

Figure 15-3 shows you the Bluetooth icon in the Quick Settings. Tap this icon to activate the signal. If you're not using a Bluetooth device, be sure to turn off the connection so as not to waste battery power.

Bluetooth

FIGURE 15-3

Bluetooth headphones, keyboards, and speakers are easy to connect (or, *pair*), and they maintain a stable connection as long as you have a line of sight between devices. Here's how to pair them to your phone:

1. Check the manufacturer's instructions to turn on the Bluetooth signal from the device you want to pair with your phone.

2. Swipe down from the top of the screen to reach the Quick Settings area.

3. Tap and hold the Bluetooth symbol to activate your phone's receiving signal.

4. A screen should show up with the Pair New Device link — tap the link. If you don't see the words *Pair New Device*, check under the Available Devices link, or tap the More link.

5. You should see a screen listing available Bluetooth devices within range of your phone. Tap the name of the device you want to pair.

6. You may see device-specific instructions on the screen. Just follow the prompts. Your devices are then paired and ready to use!

Find Favorite and Fun Apps — a Consensus

Tap the icon for the preinstalled Play Store to enter a world of fun, education, and entertainment. I've crowdsourced my friends for their favorite apps and picked my own favorites as well. Here's the consensus (from this polling) of tried-and-true apps that are easy to use and quite popular.

Note: In my unofficial research, it seems that iPhone users mostly use Apple apps, even though others are available to them. Android users use both Google apps *and* a mishmash of other developer's apps.

Personally, I like variety and new ideas.

Radio, podcasts, and music

You can find plenty of audio entertainment by using your phone with these apps:

» **IHeartRadio** is my favorite for entertainment — I can listen to local radio in almost any part of the United States. No matter how far I travel, I can listen in real time to any of 860 live broadcasts across the US. The list of stations is here www.iheartmedia.com/stations. You can also stream unlimited podcasts and music. Once the app learns your music preferences, it builds playlists for you automatically. Notice the "Your Stations" and "Your Podcasts" headings in **Figure 15-4**. The app is free, but you see ads when you use it. It also offers paid tiers that give you even more access.

» **Alexa** (from Amazon Echo) can be pretty entertaining. You can check out podcasts and news reports from major networks by enabling each provider's skill in the app. (Alexa *skills* are like apps) I've found that many of the stations I listen to on my Echo device come to Alexa from IHeartRadio, but when you have an Echo device at home, the app is more versatile.

» **Pandora** started as a free personalized music app, and it still does an excellent job of customizing the songs you hear. Most people don't realize that Pandora is owned by Sirius XM, so you can just imagine the breadth of content available. You can listen without registering, but registration enables you to create your own customized music stations. *FYI:* Your Pandora registration is public by default, but you can choose for it to be private. Do that in the account settings.

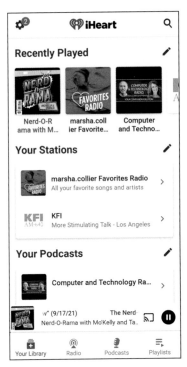

FIGURE 15-4

Video apps

You may think that YouTube is the go-to app for all things video, but now it's so much more. It's a Google property, and in addition to how-to-fix-your-toilet videos, you can now find streaming music, streaming gaming, and classic TV. A paid tier even delivers content to you ad-free.

If you have subscription TV or subscribe to all the major networks, you can watch them on not only the big screen but also your phone, using the networks' individual apps. Watching classic or live TV on your phone can sure help pass the time in the doctor's waiting room (don't forget the earbuds).

Travel

I don't keep travel apps on my phone all the time, but I find that installing these apps for periods when I'm traveling makes my life much easier. When I return home, I uninstall them until next time.

An app from the airline you're flying on gives you a boarding pass to make the process easier. Also, if a travel delay happens, you receive a notification from the app. If you need to make changes in your itinerary, you can do that from the app — on the spot.

TIP

Before you board an airplane, be sure to join the airline's frequent flyer program. You may not consider yourself a frequent flyer, but those miles can add up, and you can use them for extra perks when flying. You can track your points or miles in the app or on the airline's website.

Many hotel chains offer their own apps. I've found using these quite helpful — you can often check in on the app, request assistance from the front desk, or gain access to hotel amenities. In a foreign country, you can look up the hotel's address in the app and give it to a taxi or rideshare driver.

The MyTSA app is a splendid traveling companion; it contains a wealth of information that even the most experienced traveler can benefit from. As shown in **Figure 15-5,** you have access to checkpoint wait times by airport, FAA flight delays, items that you can (or can't) bring in carry-on or checked baggage, and more. If you're a member of TSA PreCheck®, just add your known traveler number (KTN) so that you can have that number in your records to copy on to any airline reservation you make.

Games

Many of your favorite games are available from the Play Store. You can probably find multiple apps for each game. How do you decide which app to choose? In this case, I suggest that you read the reviews. An app that stands above the others will soon be highly rated. Also, read the description to see whether ads are shown during the game. Reviews let you know whether the number of ads shown during game play is excessive.

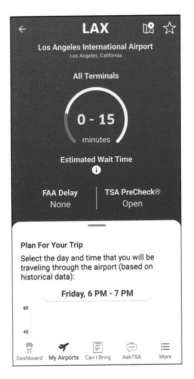

FIGURE 15-5

I download only a few apps at a time and then uninstall and find new ones. The only standard apps I keep all the time are Tetris and Sudoku. My husband is a fan of Solitaire and Backgammon.

If you're into playing Bridge, the website at www.bridgebase.com is the largest Bridge community in the world. You can often find up to 10,000 people playing online! This website's app for Android users, Bridge Base Online, has games for players from beginner to experienced. You can play against artificial intelligence or other humans in cyberspace.

For just plain entertainment value, it's tough to beat the Jeopardy and Wheel of Fortune apps.

FIND LONG-LOST ANDROID APPS

In the Play Store on your device, you have access to any app you have ever down-loaded, as long as you've used an Android device.

I noticed recently that a specific Sudoku app I used on my Android tablet is no longer available in the Play Store. Here are the steps for recovering an app you owned previously:

1. Tap to open the Play Store, and then tap your profile picture in the upper right corner.

2. Tap the Manage Apps and Device link. On the next screen, tap the Manage tab.

3. Look for the word *Installed* on a button near the top. Tap the button, and you see a menu pop up (on the left in the sidebar figure) with the options to see a list of apps Installed (on this device) or apps Not Installed.

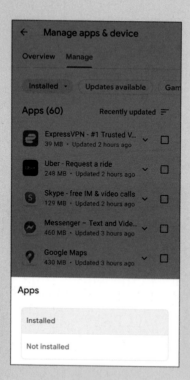 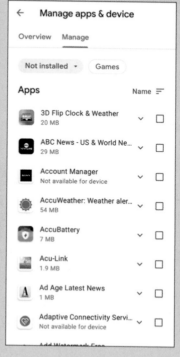

4. When you tap the Not Installed button, you can walk down memory lane and see all the old Android apps you own (on the right in the sidebar figure) — from the beginning of your Android journey.

5. If the list isn't in alphabetical order, tap the three lines on the right to sort by name.

6. When you find the app you want, tap to put a check mark in the check box to the right of the app; then tap the Download arrow at the top of the screen to download the long-lost app. The Play Store will let you know if that app is no longer compatible with your current device.

WARNING

When you download free games, you share data with the game developer — and not only developers but also organizations, businesses, insurance companies, and medical suppliers that offer games to their members. Be sure to read the privacy policy to see what happens with your data and whether it's shared or sold.

News

Google News is another of my favorite apps — I use it every morning to read the news with my breakfast and coffee. The app is a news aggregator that finds news stories for you based on your favorite topics and where you live. It updates the news feed throughout the day, so the news is fresh.

At the bottom of the screen (or in a pop-up menu), tap the icons to go to stories tailored for you, new headlines, stories you're following, or newsstand. **Figure 15-6** shows the options on two phones.

If you have digital subscriptions to newspapers, you *can* find an app for each one. If you subscribe to more than one newspaper, switching between apps can make for some convoluted daily reading. After you sign in as a Google News subscriber, the app figures out which newspapers you're subscribed to and serves up stories in your feed.

FIGURE 15-6

Books

Reading a book on a small phone screen is a challenge for me, but many people do it successfully. (I need to use a small, cheap tablet instead.) If you own a device capable of reading books or buy reading material from online merchants, an app is free to download books and magazines to your device.

» Amazon's **Kindle** and Barnes & Noble's **Nook** apps can both be downloaded for free from the Play Store. These help you buy (and read) books from merchants' websites.

» With the free **Libby** app from Overdrive, you need to have a library card, and with the card as identification, you can then locate your local library. Over 90 percent of public libraries in North America use OverDrive, and Libby can be found in 78 countries worldwide. The app enables you to virtually check out books, magazines and newspapers, and audiobooks for free. You can search for the books you want to read and download them to your phone or — to save storage space — stream them as you read from the cloud to your device.

Engage Social Media

Let's not forget social media apps — like it or not, social media has become ingrained in everyone's lives. This list briefly describes the major sites:

» **Facebook** is the best way to connect with old friends.

» **Instagram** (owned by Facebook) is the home of an endless display of pictures. You can follow your friends and their everyday photo moments or check out hashtags that follow your interests.

» **LinkedIn** is useful if you're still working. Many consultants find jobs just by participating on the site.

» At **TikTok,** a user-generated video website, people post the strangest things. Some say it's an amusing way to spend an hour.

» **Twitter** is the place to find breaking news and interesting opinions.

REMEMBER

In any social media feed, you find an endless supply of hashtags — a *hashtag* is a topic, word, or phrase preceded by the pound sign (#). Click a hashtag to find many posts on the same subject. For example, tap #cats to see photos, stories, and much more about our furry friends.

If you want to find out more about social networking, you can enjoy another one of my books — *Facebook, Twitter, & Instagram For Seniors For Dummies* (Wiley).

Try Out the Android Accessibility Suite

In the Accessibility Suite — one of the best-kept secrets of Android smartphones — you find lots of adjustments that can make your phone easier to use. You won't find an icon for this tool in the app list, though, unless you put it there. Seems counterintuitive, doesn't it? We can fix that right now.

To find the Accessibility Suite of tools, swipe downward on the Home screen and tap the Settings cog in the upper right corner. In the main settings area, type the search term *Accessibility* in the search box at the top of the screen.

On some phones, you find almost four pages of options to enhance your phone experience. **Figure 15-7** shows just two screens' worth of options that are available on the OnePlus 9 Pro 5G.

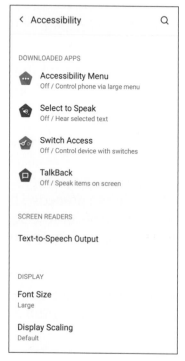
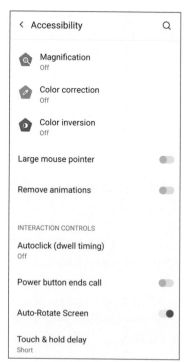

FIGURE 15-7

Because you can find many options, I recommend that you review them yourself. Let me give you some of my favorite highlights:

» **Accessibility Menu:** Activating the Accessibility button in Advanced settings installs a small human figure icon on the far right of the bottom screen navigation menu that appears on every screen on your phone. The icon is also on the Home screen. No matter what screen you are on, when you tap the small human figure, an Accessibility menu (see an example in **Figure 15-8**) appears.

Rather than fiddle with the buttons or other settings on the phone, you can tap options on this menu to lock the phone, control the volume, and power down the phone. You can also control the brightness, take screen shots, and access all the accessibility settings.

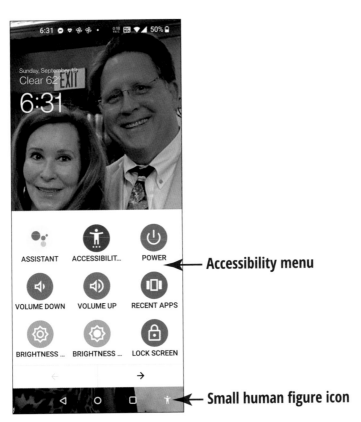

Accessibility menu

Small human figure icon

FIGURE 15-8

» **Select to Speak:** Tap some text on the screen, and it's read aloud to you.

» **TalkBack:** When the TalkBack screen reader is on, you can speak passwords, control your device with gestures, and type using the onscreen Braille keyboard. Your phone also reads onscreen text to you. Tapping the TalkBack link walks you through all the options and gives you step-by-step instructions on how to set things up.

Other excellent tools can make the mouse pointer larger, allow you to magnify any screen on your phone, or deliver closed captioning for Live Caption videos. You can also make your phone hearing-aid-compatible.

Accessibility is an ever-growing area of the Android operating system that's updated regularly and worth checking out in depth.

5

Android Today and Tomorrow

IN THIS PART . . .

Tweaking your phone's usability

Connecting safely at home and in public

Establishing a recharging strategy

Activating emergency features

Sprucing up the Home screen

Anticipating the Android 12 user experience

Chapter **16**

Marsha's MUST-DO Things for Your Phone

Acquiring a new phone and all its accoutrements is an exciting experience. You have your new technological wonder and can enjoy looking at it and making it your own.

In this chapter, I share a few items you might like to replicate on your phone, whether it's brand-new and needs tweaking or it's your trusty old Android that you're sprucing up.

If you have an older phone, you may not have noticed new features and icons that popped up during updates or app installations. I've fallen into that category, so in this chapter I outline some methods for keeping the Home screen clean and sharp-looking, too.

Here's to making your phone your very own, most useful tool!

Make the Orientation Decision

Deciding about the orientation of your phone's screen may seem too simple, but I'm a fan of using my phone in either vertical (Portrait) mode or horizontal (Landscape) mode depending on what I'm doing. Whenever I buy a new phone, it takes me awhile to figure out the orientation setting (I've even thought my phone was defective) — until I finally remember that some phones start out locked in the vertical position. This situation is officially called *orientation lock.*

Sometimes you can't get the Home screen to rotate — even if the rest of your screens do. Chapter 13 has the solution. (Pssst. . .it's in Home Screen settings.)

If you find yourself in this position (it can't be only me), swipe downward twice from the Home screen to access the Quick Settings window shade. Then look for a control labeled Auto Rotate, as shown in **Figure 16-1.** When you tap to toggle on that control, your phone's screen reacts automatically as you turn it from vertical to horizontal viewing.

If your phone has no Auto Rotate control on the Quick Settings window shade, you can also find it by searching for *accessibility settings* after tapping the main Settings cog.

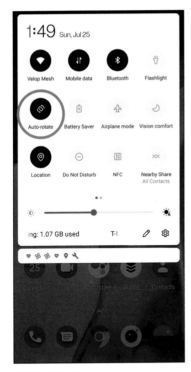

FIGURE 16-1

Configure Do Not Disturb

Few annoyances are worse than receiving a robocall at 2:30 A.M. when you're deep in sleep. Even if you haven't received one, know that the day will come. It's inevitable.

With your phone's Do Not Disturb feature, you can shut off your phone between the hours of _____ and _____. I've found that these <fill-in-the-blank> hours vary for everyone, so let's say between the hours of 10 P.M. and 9 A.M., for example.

But as a family member, you might worry that you won't be able to hear an emergency call that you *need* to answer during the specified hours. I say: No worries. You can select members of your contacts list whose numbers will ring through the Do Not Disturb feature. You also find a setting that allows a number to ring through if someone calls repeatedly within a short period.

Visit Chapter 13 for step-by-step instructions on how to set up the Do Not Disturb feature.

Use Your Home Wi-Fi for Calls and Browsing

When you're at home, be sure to connect your phone to your home Wi-Fi. Even if you're on an unlimited plan, Wi-Fi calling can be beneficial. Sometimes (like at my house), my carrier's signal isn't always the strongest — the signal can even vary from room to room. Setting up Wi-Fi calling ensures that I can complete a phone call even when walking from room to room. Also, some app settings will update, upload, or configure only when connected to Wi-Fi and not when using mobile data.

When you have this home Wi-Fi connection set up, your phone routes the call over the best-quality network, whether it's your home Wi-Fi or the carrier's signal. This choice of networks makes for a better calling experience, though I do yearn for the days of those perfect landline connections.

This Wi-Fi connection isn't automatically set up on all phones — you may have to make the connection yourself. (Your carrier might not even offer the Wi-Fi calling option.) The procedure may vary from brand to brand, but here are the usual steps:

1. Swipe downward on your phone's Home screen, tap the main Settings cog, and then tap Connections. On the Connections screen, toggle the Wi-Fi Calling option to On. You're done.

2. If you don't find the Wi-Fi Calling option as in Step 1, swipe downward on the Home screen and tap the Settings cog. Then type *wi-fi calling* into the search box next to the Magnifying Glass icon on the Settings main page. Follow the search results and toggle the Wi-Fi Calling option to On.

3. If Steps 1 and 2 don't work, search for *wi-fi and network* (or similar wording); then look for *SIM and network*. This strategy should bring you to a screen with the Wi-Fi Calling option; toggle it to On.

WARNING

If you're on a metered cellular plan, Wi-Fi calling may still charge your account by the minute and cost you extra (especially with International calls). Check with your carrier to be sure.

Practice Safety When Using Public Connections

Have you ever been away from home when your cellular connection couldn't connect? Yes, that's happened to me, too — usually, at the airport, a hotel, a large event, or a conference. In this type of situation, you're grateful to discover free or reasonably priced Wi-Fi available.

Use a VPN to secure public Wi-Fi connections

Fair warning: Using public Wi-Fi (even at a reputable coffee shop or museum) can leave your connection open to snoopers and others who can enter your connection and steal information from your device. These ne'er-do-wells can go as far as setting up dummy Wi-Fi connections that connect you to the Internet and deposit malware on your phone.

What to do to prevent such invasion? Consider subscribing to a virtual private network (VPN) service to route your Internet traffic. VPNs are commonly used in business to protect private communications. Many people I know even use VPNs for their connections from home, for privacy's sake.

The VPN subscription works by connecting you to the Internet by way of a portal or a VPN app, and your identity stays private. You're connecting to the Internet via the VPN server, not your device. When you use a VPN, no one knows who you are or where you go on the Internet — not even your phone provider. Your provider only knows that you're connecting to a server.

You can find many VPN options to choose from, but you get what you pay for in this area, and free isn't a fully secure option. If you want to use a VPN for a trip or special event, you can pay for just a month of service. My husband and I use a VPN at home too — and subscribe by the year. Our subscription covers five devices at a time.

I can recommend the VPN service I've used for years: ExpressVPN, at www.expressvpn.com, has been ranked the safest by top publications. **Figure 16-2** shows a phone's screen when connected to a VPN by way of the ExpressVPN app.

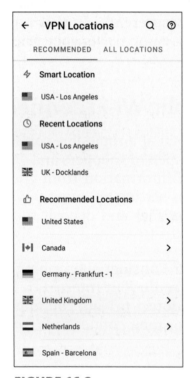 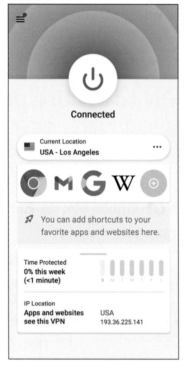

FIGURE 16-2

WHEN THERE'S LITTLE CELLULAR SIGNAL OR WI-FI

Having a weak or erratic cellular signal is a sticky situation, and it usually occurs when you're on vacation in another country or visiting a remote area. Because I *always* need to be connected, I sought out a solution.

My husband and I planned a winter trip to Alaska that took us 100 miles north of the Arctic Circle, and I doubted that I would find a signal out in the middle of nowhere. Although the Skyroam customer service team doubted my strategy, I gambled and bought a small device the size of a hockey puck: the Skyroam Solis hotspot. I knew that truckers drive routes right up the middle of Alaska — and have connection — so there must be a signal somewhere. There was, and thanks to my Solis hotspot, I had a 4G LTE signal on a dogsled.

Skyroam partners with cellular providers all over the world (and 130 countries) to share their signals. It won't work on a cruise ship or on a plane, but it has never let me down when traveling. You can buy data by the gigabyte or by unlimited day pass. Look into this reasonably priced travel solution at www.skyroam.com.

Beware of charging a phone from a public port

Surprisingly, even using a public Wi-Fi charger can be perilous. Think twice before you use one. The practice, called *juice jacking*, enables a hacker to install malware on your phone when you plug into a public charging station. With this practice, hackers can embed malicious code into USB sockets (the charging ports) that can transmit your phone data to an attacker's servers. Even when you disconnect from the outlet, the code now resides in your phone!

You can plug a lightweight portable power bank into a public station with no risk. Power banks are simply batteries and cannot be hacked. When I'm on a trip, I plug in to charge my external batteries wherever and however I can, for as long as I can.

If you're going to be out and about and run the risk of draining your phone's battery, carry a small portable charger in your pocket or purse. You charge the portable chargers at home, and they, in turn, can safely charge your phone in an emergency. Better to be safe than sorry.

You can buy a pocket-size 10,000 mAh (milli-amps per hour) power bank online for about $35. (Today's phone batteries have a capacity of 4,000 mAh in storage.) **Figure 16-3** shows a power bank connected to an Android phone.

Photo courtesy of Boosa Tech

FIGURE 16-3

If you're away from home and totally stuck with no phone battery power, use your phone's own power cable and charging block plugged directly into a power outlet.

Secure Your Power Cables

Power cables are a real consideration. I once thought that a cable is a cable is a cable, but nope — cables differ in quality and can be rated for data and/or power. Cheap, thin, flimsy cables can become a hazard to your phone. Also, you should keep the original power cables and blocks together with their related devices.

The power block (the gizmo your phone's power cable plugs into) varies from device to device. I studied the information printed in tiny letters on the bucketful of blocks I have at home and found that their voltage varies considerably. Also, some smartphone manufacturers embed chips to prevent overcharging.

Turn on Battery Saver mode in your phone's Quick Settings if your phone's percentage of charge sinks too low. Extend your phone battery's longevity by not letting the charge run down lower than 20 percent or charge to over 80 percent. The lower the battery percentage, the lower its voltage. The higher the battery percentage, the greater chance that the internal circuits can overheat.

Take a Sharpie and write directly on the charging block the name of the device it came supplied with. If the charging block is black, use a cheap label maker and stick the device name right on the block. I use the label maker to wrap an identifying label around the cable, too. Yes, it sounds a little crazy, but in this world of expensive electronic devices, you want them to last as long as possible.

Set Up Emergency Call and SOS

In a true emergency, you may not have time to call 911 or even open your phone to dial an emergency call. Here's where emergency SOS comes in.

As with all procedures on Android phones, you can find several ways to activate an emergency beacon to send your location to emergency services. This notification (which is different from the emergency information I introduce in Chapter 4) can engage when you need urgent help and can't make a regular call.

Establishing SOS messaging

Enabling an Emergency mode SOS has pros and cons, and it's up to you to decide whether to activate this feature:

» **Pro:** A stranger who picks up your phone will be unable to access your personal and medical information. Instead, Emergency mode SOS can set off an alarm, send messages to specified contacts, and/or contact emergency services to supply your location.

» **Con:** If you keep your phone in a pocket or a purse, you can activate Emergency mode accidentally if your phone hits against another item in your pocket or purse three (or five) times in rapid succession. This situation is unlikely to happen.

To set up SOS messages, follow these steps:

1. Swipe downward on the Home screen and tap the main Settings cog. Scroll down and tap Advanced Features on a Samsung phone and search Emergency Rescue (or Emergency SOS) on other Android devices.

2. On the Advanced Features screen on Samsung, tap the Send SOS Messages option and on others tap Emergency Rescue. Then toggle this to the On position (or tap for more options on other Android brands).

3. On the resulting screen, similar to the one shown in **Figure 16-4,** read the info and tap Continue. Then follow the prompts to grant permissions, specify contacts, set the number of times the power button is pressed to activate SOS, and even add a photo or recorded message to send when SOS is activated.

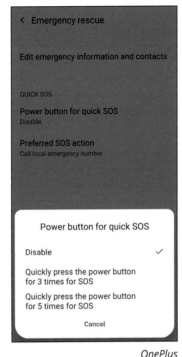

Samsung   OnePlus

FIGURE 16-4

REMEMBER

Your phone must have an active SIM card for the emergency SOS messaging to work. The messaging repeats periodically for 24 hours.

Activating Emergency mode

You can access your phone's Emergency mode, which helps (among other benefits) to save on battery power, by holding down the power button to access the Power Off or Restart screen. (On some Samsung phones, you access this screen by holding down the lower volume rocker and the Bixby button together.) You should also see the Emergency Mode icon. Tap that icon and you see a page where you can make an emergency call and add contacts or SOS features.

Activate Emergency mode if you feel that your phone's battery power is too low and you're worried that you might lose contact before you can recharge it.

On the OnePlus 9 Pro, shown in **Figure 16-5,** you see how the phone counts down 3-2-1 on the screen before it makes the emergency call. I can't guarantee what happens next because I bailed out by tapping the Cancel (X) icon.

FIGURE 16-5

Holding down the power key and then tapping Emergency on a Google Pixel phone brings you to the 911 call page, as shown in **Figure 16-6.**

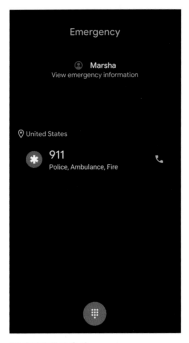

FIGURE 16-6

If you have a Samsung S10+-style phone, you can tap Emergency Mode on the shutdown screen. **Figure 16-7** (on the left) shows how the resulting screen looks. The right side of the figure shows what the Samsung S10+ looks like when it converts to a low-power state after you tap Turn On to engage Emergency mode.

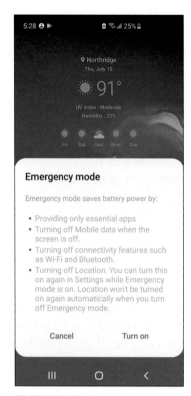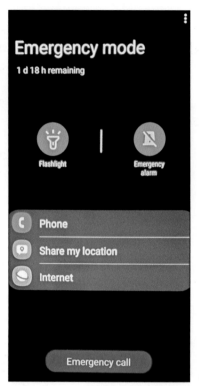

FIGURE 16-7

Managing Emergency mode

Tapping the three dots in the upper right corner of the active Emergency Mode screen produces a drop-down menu. These options appear:

» **Turn Off Emergency Mode:** When you want to leave Emergency mode, tap here.

» **Edit:** Tap to add an app such as Maps or email to the Emergency Mode screen.

» **Emergency Contacts:** Tap this option to open the Contacts page, showing people you've previously selected by adding a star to their names in the Contacts app. You can just tap to call them.

At the bottom of the Emergency Mode screen, you see the red Emergency Call button. Tap the button and then indicate whether you want to call 911 with a yes-or-no response.

REMEMBER

After you tap YES to call 911, the screen shows a countdown. If calling was a mistake, you can stop the process by tapping Cancel.

Most Android phones use Google's Emergency Location Services (ELS). This service activates only when you press the power button to contact SOS. Your location is computed on your phone and then sent securely to emergency services. (Google does not receive this information.)

TECHNICAL STUFF

The Google ELS (Emergency Location Service) is built into any handset running Android version 4.0 or later. It is widely used in Europe, but right now in the United States, Only T-Mobile enables the service. ELS uses GPS, Wi-Fi, mobile networks, and sensors to identify your location, both indoors and out, with incredible accuracy. For those who want to know more about how this works, visit the page where Google explains the science at `crisisresponse. google/emergencylocationservice/how-it-works`.

Manage Home Screen App Shortcuts

Something I love about Android is that the screens usually appear less cluttered. But the bad news is that, when you install Android apps, they plop a shortcut icon on your Home screen pages — which adds to the clutter. The apps don't just fill one Home screen page and then add another — instead, the shortcut icons randomly appear on new pages of the Home screen. You notice the increased clutter when you swipe left and see these random screens with random shortcuts.

I'm not alone in craving less clutter. Many people prefer keeping a simple, clean Home screen and accessing their apps from the app screens. Swipe upward on the Home screen to access the Apps drawer with alphabetically listed app shortcuts.

TIP

If the apps in your Apps drawer aren't in alphabetical order, tap the three dots on the upper right in the Apps drawer and tap Sort. You'll have the option to arrange them alphabetically.

Here are some steps for managing the shortcuts (the clutter) on your home screen pages:

1. Remove shortcuts that already appear on your phone's Home screens. As you set up your phone and add your favorite apps, you might see that shortcut icons (to launch the apps) have magically appeared on the Home screen.

   To remove the shortcut, long-press the app's icon, and a menu appears, as shown in **Figure 16-8**.

FIGURE 16-8

Depending on the app and on your phone's manufacturer, you may have these options:

- **Remove:** Tap to remove the shortcut icon. The app still appears and functions from the app screen.

- **Uninstall:** Tap this option to completely uninstall the app from your phone. If you ever want to use the app again, go to the Play Store and download it again.

- **Edit:** This option lets you change the wording (the name) that appears below the app's icon.

- **App Info (or *I* in a circle on Samsung phones):** Tap this option to see a screen that lists the options for information about the app and how your phone uses it (see **Figure 16-9**). Tap any of the listings to access the information. You can open, uninstall, or stop the app from the App Info screen as well.

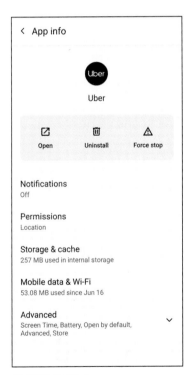

FIGURE 16-9

If an app starts acting unpredictably, tap Force Stop to close it. To be on the safe side, tap Clear Cache, which clears the processes. You may need to access Clear Cache from the Storage option on the App Info screen.

Depending on your phone and on how it lays out shortcuts, you might want to keep a second Home screen with folders (see the next section, "Make Folders of Apps").

2. Add a shortcut to the Home screen. If you have an app that you want to appear front-and-center on the Home screen, follow these steps:

   • Go to the app screen and select the app by long-pressing its icon. (Long-pressing certain app icons makes a menu appear, but ignore it for now.)

   • Keeping your finger on the app, begin to drag it. Almost like magic, the Home Screen page appears.

   • Lift your finger, and the icon drops to the screen.

   • To move a shortcut icon to the main Home screen (if it's not already there), select it and move it again, as you just did.

3. Prevent shortcuts from ever again being placed on the Home screen. This is easy!

   • Long-press a blank spot on the Home screen. The screen transforms, as shown in **Figure 16-10.**

   • Tap Settings to get to the Home Screen Settings page. Scroll to find an On-Off toggle that enables adding new apps to the Home screen.

   • Tap to toggle off this setting, and you never again have to deal with apps installing themselves on your phone's Home screens.

 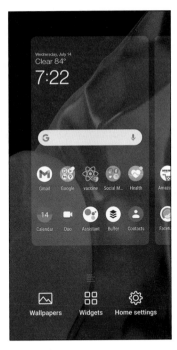

FIGURE 16-10

Make Folders of Apps

As you may have noticed from the figures in this book, I'm a fan of grouping together certain apps into folders for organization and a less cluttered screen appearance. They may be social media, travel, or shopping apps — how you group apps is your call.

To make a folder, select an app by long-pressing its icon and moving it over another app icon. The two form a folder into which you can add other apps. In **Figure 16-11,** you can see that I added the Instagram app to the Social folder.

After you create a folder, you can tap *Folder Name* and type the name you want for the folder. On the Pixel series of phones, folders receive a temporary name.

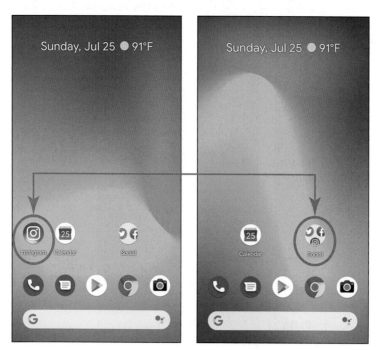

FIGURE 16-11

Chapter **17**

Android 12 and Beyond: The OS Evolution

As noted in the book's introduction, I wrote this book using the Android 11 operating system (OS) to explain all the ins and outs of Android smartphones. To get a better look at Android 12 (the next version of the OS), I joined the beta preview group for the OnePlus 9 5G and the Pixel 3XL, and I worked with a technically minded friend to take this chapter's screen shots on a Pixel.

Exploring beta previews is not for the faint of heart. I don't recommend that you do so unless you're a closet tech-nerd who remains unfazed when processes break down (and it's guaranteed to happen).

Ask yourself this question: How much more do you want your phone to do? I honestly don't want much more from mine. (I have to work hard to take a bad picture on a smartphone these days). All I want is for my phones to continue getting easier to use.

Update after update happens on any Android phone, and over the course of a year, small differences can offer insight into operating system changes (if you're paying attention). While writing this book,

I dug deep into the evolution of the Android operating system, and now I want to share my conclusion: *You needn't fear the future.*

In this chapter, I offer a glimpse of the small changes coming in Android 12 that may lead to an even better user experience.

Find Helpful New Features in Android 12

During Android 12 beta testing, I watched the product's look-and-feel morph into the final version (or close to it). One thing became clear: The operating system you know from this book (Android 11) isn't changing much.

Make purchasing faster with GPay

The basic processes of the Android OS that you (no doubt) love are all there; some are more refined and easier to use. For example, it seems (as of now) that going to the Quick Settings (by swiping downward on the Home screen twice in a row) opens the door to Google Pay (GPay). In GPay, if you want, you can enter your credit or business loyalty cards to use directly from your phone.

After you tap the GPay icon, tap Add a Payment Method (as shown in **Figure 17-1**), where you may see sample cards. Tap the card you want to add, and GPay then knows whether you need to type in the card info or simply scan the card. Follow the instructions onscreen; I don't offer specific instructions, because they may change when the final Android 12 OS is released.

After you load your cards, you can use them for a purchase by either tapping a Near Field Communications (NFC) device or having the vendor scan your phone.

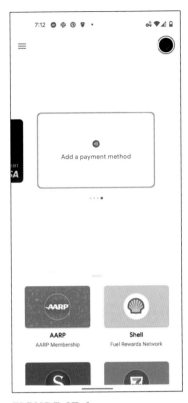
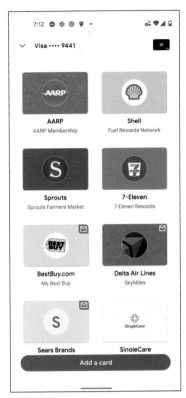

FIGURE 17-1

Keep in mind that the GPay example is conjecture at this point, and features may change before Android 12 is released. I don't offer more specific examples, for that reason.

Poke around in Android 12

As you discover in this book, tapping and poking around are the best ways to find and learn things on your phone. Try out apps and settings to see whether you like them. My best recommendation to you is to search in the main Settings area. Look for something new when Android 12 appears and give it a whirl.

As more developments happen with the Android 12 OS evolution, I plan to share information about the changes on my blog (mcollier.blogspot.com) and Twitter account (@MarshaCollier).

Meet the Soothing Android 12 User Experience

As I asked in this chapter's introduction, how much more do you want your phone to do? I want my phone to be easy to use, and I also want it to look nice. One advantage of Android is not having to deal with cluttered screens. The Home and Lock screens on an Android smartphone can be works of art.

That's why Android 12 presents the *Material You* interface. This interface isn't a *skin* (the manufacturer's look and feel), as in branded phones, but is built into the architecture of your device. The goal of this interface is to make all the elements on your phone look like they belong together. The dozens of app icons on your phone may still have their own designs, but with Material You, they have a commonality that makes your phone seem more artful and appealing.

Figure 17-2 shows an image selected from Pixel wallpapers for the Home screen background. After choosing the image, I applied a theme that directed Android 12 to build interface elements from the colors in the image. On the screen shown on the left, notice that the initial app icons look pretty much as they do in Android 11. After applying the Android 12 theme, the icons look like they belong together and coordinate with the background image. Very cool, yes?

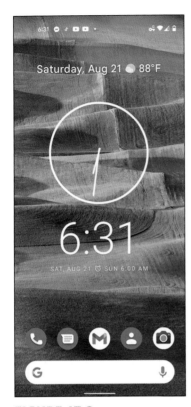

FIGURE 17-2

Android then applies this theme-color palette to other functions — for example, long-pressing an app icon on the Home screen. The flyout menu now coordinates with the look-and-feel of the phone. **Figure 17-3** shows two sample menus.

These sample screens from Android 12 suggest that the goal is to make your phone feel less like a busy, complex computer and more like a trusted tool you'll want to use every single day.

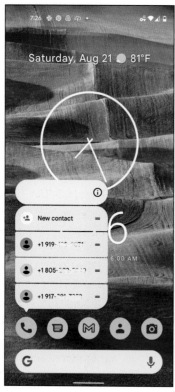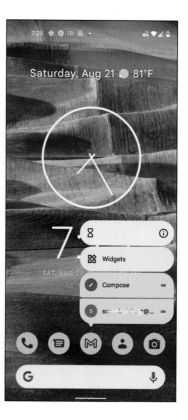

FIGURE 17-3

Be Ready for the Future

We live in exciting times, and being aware of ever-changing technology is a big step toward enjoying the related advances. After all, nations sent men to the moon using less computing power than you have in your smartphone.

Don't be afraid to try out all the different apps and settings, and be happy when you accomplish assorted tasks on your phone. Share your knowledge with your friends! Keep in mind that it's difficult to end up in real trouble — just keep track of the steps you take when trying something new. (I take screen shots and delete them after I'm successful.) I hope you learn a lot, or at least become more confident by reading this book. And I hope to regain your readership, maybe when Android 14 rolls around?

Index

J

jack connections, 266

Jeopardy game app, 271

juice jacking, 287

junk mail, 190

K

Keep app, 260

Keep Contacts Up to Date, Google Contacts app, 161–164

keyboards

diacritical characters on, 101–102

emojis on, 102–103

features of, 89–93

Gboard keyboard

diacritical characters, 101–102

emojis, 97, 102–103

features of, 95–97

printing on, 105–108

QWERTY keyboard, 174

revising text with, 104–105

spell-checking with, 100

voice typing with, 99–100

high contrast, 92–93

Microsoft SwiftKey, 99

overview, 94–95

printing on, 105–108

revising text with, 104–105

Samsung, 97–98

searching in settings for, 93–94

selecting, 88–89

spell-checking with, 100

themes, 91

voice typing with, 99–100

Kindle app, 274

known traveler number (KTN), adding to MyTSA app, 270

L

lag time (latency), 16

Landscape mode

adjusting screen to, 210

orientation setting of, 282

for Samsung keyboards, 97

latency (lag time), 16

LCD (liquid crystal display), 24

Lens app, 212–214, 219

Libby app, 274–275

Light mode, 92

LinkedIn app, 275

links, copying, 187

liquid crystal display (LCD), 24

Location Quick Setting, 232

locations

previewing on maps, 257

setting up for calendar appointments on Google Maps, 255–257

sharing, 178

lock screens

overview, 54

placing owner information on, 59–62

process of, 56–58

type options for, 54–56

locked phones, 31

long-pressing gesture, 77, 251

LTE (Long Term Evolution), 16

M

macros, for taking photos, 212

magic wand, adding filters to photos with, 212

mAh (milli-amps per hour), 288

manufacturers, tech support though, 15

Maps app

Directions screen options on, 257–258

integrating with Google Calendar, 243

interactive, 178

overview, 255

setting calendar appointments up with locations, 255–257

Material You interface, 304

MB (megabytes), 13–14

media apps, downloading, 82–83

media button, 74, 121

medical information, 63–65

megabytes (MB), 13–14

megapixels (MP), 206–207

menus

Accessibility Suite, 277

for cameras, 214

Hamburger, 190, 191

main, 190–191

scrolling, accessing camera features on, 214–215

messages

email, 185

text, 157

dictated, 99–100, 180–181

editing, 183

emergency, 289

Google Messages, 173–176

scheduling messages for later delivery, 183–184

sharing, 184–185

messaging protocols, 168

metadata, 217

microphone icons, 180–181

Microsoft, involvement with Samsung, 8–9

Microsoft patent licensing, 9

About the Author

Marsha Collier is a best-selling author, speaker, and radio host covering the topics of e-commerce, customer service, and social media. She is also one of the foremost eBay experts and educators worldwide. She has sold over one million books and has published 48 (plus) titles in the For Dummies series (John Wiley & Sons, Inc.). Marsha's titles covering eBay and social media include the best-selling *eBay For Dummies* and *Facebook, Twitter, and Instagram For Seniors For Dummies*.

Making online customers happy and winning new customers are her goals. Marsha's books *Ultimate Online Customer Service Guide: How to Connect with your Customers to Sell More!* and *Social Media Commerce For Dummies* show how social media is an unparalleled vehicle for connecting with an unlimited customer base.

Marsha authors the popular blog Marsha Collier's Musings for small business and operates Cool eBay Tools, a Web resource for online sellers. For more tech tips, listen to her podcast Computer and Technology Radio on your favorite podcast network.

Marsha is recognized by major publications as a social media and technology Power Influencer; you can find her full bio on Wikipedia and LinkedIn.

Dedication

I wrote this book for everyone "over a certain age" who likes to be fluent and self-sufficient with personal technology devices.

Author's Acknowledgments

This book wouldn't be here without the sage advice (and inspiration) from my Project Editor, Leah Michael. She tried to retire, but this side gig of hers was a blessing to me. TJ McCue? I love your dedication to technology peppered with a great sense of humor. Thank you for partnering with me again. I'm pretty sure that Strunk & White have nothing over Becky Whitney. She's an eagle eye when it comes to editing. Thanks to Steve Hayes for persuading me to write this book. I really enjoyed this project, and it's thanks to this wonderful team.

Most of all, thanks to my husband Curt Buthman who listened to me drone on about Android (even though he's an iPhone user). He offered input (text messages and such) for the book's screen shots and supported me with everything I needed. Most of all, he put up with my skipping dinner when I was in a writing frenzy.

My new kitten, Sammy, kept me smiling and grew up as I was writing. You can see him in some of this book's figures.

Publisher's Acknowledgments

Acquisitions Editor: Steve Hayes

Development Editor: Leah Michael

Copy Editor: Becky Whitney

Technical Editor: TJ McCue

Managing Editor: Kristie Pyles

Production Editor: Mohammed Zafar Ali

Cover Image: © Photo courtesy of Curt Buthman